THE LAYMAN does not often have the opportunity of reading a simple exposition of advanced scientific thought written by the man who did the actual creative thinking.

In this book, which is the result of a happy collaboration between the author of the theory of Relativity and one of his own co-workers in research, there will be found an easily understood but authoritative account of the growth of ideas in physical science from the earliest concepts to the more abstruse theories of modern times. The story they have to tell of this evolutionary development is one of the most fascinating that the human mind can meet with—the story of mankind's attempt to comprehend through inventive thought its own relationship to the external world.

In simple, straightforward language, avoiding all highly technical terms and mathematical formulae, the authors have traced with beautiful clarity the steps from the mechanical view of the universe invented by the classical physicists, through the decline of this mechanical view, to the more satisfactory explanations evolved by modern science.

They illustrate their more difficult points with graphic clearness by a series of diagrams, and by using comparisons with the experiences of everyday life. They make comprehensible to the average reader the whole range of evolving thought in physical science, and they explain the significance of the most important contributions since the time of Newton—the inventions of the ideas of Field, Relativity, and Quanta.

Here is the story of man's conquest of his own ignorance. To read it is to participate in one of the greatest adventures of all time — the adventure of expanding the horizon of knowledge, the adventure of man's magnificent struggle to understand the laws governing the universe in which he lives. THE PUBLISHERS

The original edition of The Evolution of Physics *was published simultaneously in England by the Cambridge University Press, in Holland by A. W. Sythoff's Uitgeversmaatschappij, N.V., and in the United States by Simon and Schuster.*

The
EVOLUTION
of PHYSICS

from
Early Concepts

to

Relativity and Quanta

by

ALBERT EINSTEIN
and LEOPOLD INFELD

A Touchstone Book
PUBLISHED BY SIMON AND SCHUSTER

ISBN 0-671-20156-5
MANUFACTURED IN THE UNITED STATES OF AMERICA
PRINTED BY MURRAY PRINTING CO., FORGE VILLAGE, MASS.

15   16   17   18   19   20

Table of Contents

List of Plates

Acknowledgments

We wish to thank all those who have so kindly helped us with the preparation of this book, in particular:

Professors A. G. Shenstone, Princeton, N. J., and St. Loria, Lwow, Poland, for photographs on plate III.

I. N. Steinberg for his drawings.

Dr. M. Phillips for reading the manuscript and for her very kind help.

A. E. and L. I.

Introduction to the New Edition

THE FIRST EDITION *of this book appeared more than twenty years ago. Since then, death has come to Einstein, its chief author and perhaps the greatest scientist and the kindest man who ever lived. Since then, too, there has been an unparalleled development of physics. It is enough to mention the progress in nuclear science and the theory of elementary particles, the exploration of cosmic space. Nevertheless, there is very little that must be changed in this book, since it deals only with the principal ideas of physics, which have remained essentially the same. As far as I can see, only a few minor corrections are necessary.*

First: the book deals with the evolution of ideas and is not a historical account. Therefore, the dates given are usually approximate and put in the form ". . . many years ago." For example, in Chapter IV, "Quanta," the section "Light Quanta" (p. 268), we wrote about Bohr: "His theory, formulated twenty-five years ago . . ." Since the book was first printed in 1938, "twenty-five years ago" means 1913, the year Bohr's paper appeared. And the reader should remember that all similar expressions relate to the year 1938.

Second: in Chapter III, "Field Relativity," the section "Ether and Motion" (p. 166), we wrote: "There is nothing irrational in either of these examples except that in both cases we should have to run with a speed of about four hundred yards per second, and we can very well imagine that further technical development will make such speeds possible." Today, everyone knows that the jet plane has already achieved supersonic speed.

Third: in the same chapter, the section "Relativity and Mechanics" (p. 195), we wrote: ". . . from hydrogen, the lightest, to uranium, the heaviest. . . ." This is no longer a true classification since uranium is no longer the heaviest element.

Fourth: again in Chapter III, the section "General Relativity and its Verification" (p. 239), we wrote on the perihelian motion of Mercury: "We see how small the effect is, and how hopeless it would be to seek it in the case of planets further removed from the sun." More recent measurements have revealed that this effect is true not only for Mercury but for other planets as well. It is very small but it seems to be in agreement with theory. Perhaps in the near future this effect can also be checked for the artificial satellites.

In Chapter IV, "Quanta," the section "Probability Waves" (p. 281), we wrote on the diffraction of single electrons: "It need hardly be mentioned that this is an idealized experiment which cannot be carried out in reality but may well be imagined." It is worth mentioning that in 1949 a Soviet physicist, Professor V. Fabrikant, and his colleagues performed an experiment in which they observed the diffraction of single electrons.

With these few changes, the book becomes up to date. I did not wish to introduce these small corrections in the *body of the text because I feel that a book written with Einstein should remain as we worked it out together. I am very happy that this book is still alive after his death, as, indeed, are all his works.*

WARSAW, OCTOBER 1960

LEOPOLD INFELD

Preface

BEFORE you begin reading, you rightly expect some simple questions to be answered. For what purpose has this book been written? Who is the imaginary reader for whom it is meant?

It is difficult to begin by answering these questions clearly and convincingly. This would be much easier, though quite superfluous, at the end of the book. We find it simpler to say just what this book does not intend to be. We have not written a textbook of physics. Here is no systematic course in elementary physical facts and theories. Our intention was rather to sketch in broad outline the attempts of the human mind to find a connection between the world of ideas and the world of phenomena. We have tried to show the active forces which compel science to invent ideas corresponding to the reality of our world. But our representation had to be simple. Through the maze of facts and concepts we had to choose some highway which seemed to us most characteristic and significant. Facts and theories not reached by this road had to be omitted. We were forced, by our general aim, to make a definite choice of facts and ideas. The importance of a problem should not be judged by the number of pages devoted to it. Some essential lines of thought have been left out, not because they seemed to us unimportant, but because they do not lie along the road we have chosen.

Whilst writing the book we had long discussions as to the characteristics of our idealized reader and worried a good deal about him. We had him making up for a complete lack of any concrete knowledge of physics and mathematics by quite a great number of virtues. We found him interested in physical and philosophical ideas and we were forced to admire the patience with which he struggled through the less interesting and more difficult passages. He realized that in order to understand any page he must have read the preceding ones carefully. He knew that a scientific book, even though popular, must not be read in the same way as a novel.

The book is a simple chat between you and us. You may find it boring or interesting, dull or exciting, but our aim will be accomplished if these pages give you some idea of the eternal struggle of the inventive human mind for a fuller understanding of the laws governing physical phenomena.

I. THE RISE
OF THE MECHANICAL VIEW

The Rise of the Mechanical View

The great mystery story ... The first clew ... Vectors ...
The riddle of motion ... One clew remains ... Is heat a
substance? ... The roller-coaster ... The rate of exchange
... The philosophical background ... The kinetic theory
of matter.

THE GREAT MYSTERY STORY

IN IMAGINATION there exists the perfect mystery story.
Such a story presents all the essential clews, and compels us to form our own theory of the case. If we
follow the plot carefully we arrive at the complete
solution for ourselves just before the author's disclo-
sure at the end of the book. The solution itself, contrary to those of inferior mysteries, does not disap-
point us; moreover, it appears at the very moment we
expect it.

Can we liken the reader of such a book to the scientists, who throughout successive generations continue
to seek solutions of the mysteries in the book of nature? The comparison is false and will have to be abandoned later, but it has a modicum of justification which
may be extended and modified to make it more appropriate to the endeavor of science to solve the mystery
of the universe.

This great mystery story is still unsolved. We cannot even be sure that it has a final solution. The reading has already given us much; it has taught us the

3

rudiments of the language of nature; it has enabled us to understand many of the clews, and has been a source of joy and excitement in the oftentimes painful advance of science. But we realize that in spite of all the volumes read and understood we are still far from a complete solution, if, indeed, such a thing exists at all. At every stage we try to find an explanation consistent with the clews already discovered. Tentatively accepted theories have explained many of the facts, but no general solution compatible with all known clews has yet been evolved. Very often a seemingly perfect theory has proved inadequate in the light of further reading. New facts appear, contradicting the theory or unexplained by it. The more we read, the more fully do we appreciate the perfect construction of the book, even though a complete solution seems to recede as we advance.

In nearly every detective novel since the admirable stories of Conan Doyle there comes a time where the investigator has collected all the facts he needs for at least some phase of his problem. These facts often seem quite strange, incoherent, and wholly unrelated. The great detective, however, realizes that no further investigation is needed at the moment, and that only pure thinking will lead to a correlation of the facts collected. So he plays his violin, or lounges in his armchair enjoying a pipe, when suddenly, by Jove, he has it! Not only does he have an explanation for the clews at hand, but he knows that certain other events must have happened. Since he now knows exactly where to look for it, he may go out, if he likes, to collect further confirmation for his theory.

The scientist reading the book of nature, if we may

be allowed to repeat the trite phrase, must find the
solution for himself, for he cannot, as impatient readers
of other stories often do, turn to the end of the book.
In our case the reader is also the investigator, seeking
to explain, at least in part, the relation of events to
their rich context. To obtain even a partial solution the
scientist must collect the unordered facts available and
make them coherent and understandable by creative
thought.

It is our aim, in the following pages, to describe in
broad outline that work of physicists which corresponds to the pure thinking of the investigator. We
shall be chiefly concerned with the role of thoughts
and ideas in the adventurous search for knowledge of
the physical world.

THE FIRST CLEW

Attempts to read the great mystery story are as old
as human thought itself. Only a little over three hundred years ago, however, did scientists begin to understand the language of the story. Since that time, the age
of Galileo and Newton, the reading has proceeded
rapidly. Techniques of investigation, systematic methods of finding and following clews, have been developed. Some of the riddles of nature have been solved
although many of the solutions have proved temporary
and superficial in the light of further research.

A most fundamental problem, for thousands of years
wholly obscured by its complications, is that of motion. All those motions we observe in nature, that of a
stone thrown into the air, a ship sailing the sea, a cart
pushed along the street, are in reality very intricate.

To understand these phenomena it is wise to begin with the simplest possible cases, and proceed gradually to the more complicated ones. Consider a body at rest, where there is no motion at all. To change the position of such a body it is necessary to exert some influence upon it, to push it or lift it, or let other bodies, such as horses or steam engines, act upon it. Our intuitive idea is that motion is connected with the acts of pushing, lifting or pulling. Repeated experience would make us risk the further statement that we must push harder if we wish to move the body faster. It seems natural to conclude that the stronger the action exerted on a body, the greater will be its speed. A four-horse carriage goes faster than a carriage drawn by only two horses. Intuition thus tells us that speed is essentially connected with action.

It is a familiar fact to readers of detective fiction that a false clew muddles the story and postpones the solution. The method of reasoning dictated by intuition was wrong and led to false ideas of motion which were held for centuries. Aristotle's great authority throughout Europe was perhaps the chief reason for the long belief in this intuitive idea. We read in the *Mechanics*, for two thousand years attributed to him:

The moving body comes to a standstill when the force which pushes it along can no longer so act as to push it.

The discovery and use of scientific reasoning by Galileo was one of the most important achievements in the history of human thought, and marks the real beginning of physics. This discovery taught us that intuitive conclusions based on immediate observation are not always to be trusted, for they sometimes lead to the wrong clews.

But where does intuition go wrong? Can it possibly be wrong to say that a carriage drawn by four horses must travel faster than one drawn by only two?

Let us examine the fundamental facts of motion more closely, starting with simple everyday experiences familiar to mankind since the beginning of civilization and gained in the hard struggle for existence.

Suppose that someone going along a level road with a pushcart suddenly stops pushing. The cart will go on moving for a short distance before coming to rest. We ask: how is it possible to increase this distance? There are various ways, such as oiling the wheels, and making the road very smooth. The more easily the wheels turn, and the smoother the road, the longer the cart will go on moving. And just what has been done by the oiling and smoothing? Only this: the external influences have been made smaller. The effect of what is called friction has been diminished, both in the wheels and between the wheels and the road. This is already a theoretical interpretation of the observable evidence, an interpretation which is, in fact, arbitrary. One significant step further and we shall have the right clew. Imagine a road perfectly smooth, and wheels with no friction at all. Then there would be nothing to stop the cart, so that it would run forever. This conclusion is reached only by thinking of an idealized experiment, which can never be actually performed, since it is impossible to eliminate all external influences. The idealized experiment shows the clew which really formed the foundation of the mechanics of motion.

Comparing the two methods of approaching the problem we can say: the intuitive idea is —— the greater the action the greater the velocity. Thus the velocity shows whether or not external forces are acting on a body. The new clew found by Galileo is: if a body is neither pushed, pulled, nor acted on in any other way, or, more briefly, if no external forces act on a body, it moves uniformly, that is, always with the same velocity along a straight line. Thus, the velocity does not show whether or not external forces are acting on a body. Galileo's conclusion, the correct one, was formulated a generation later by Newton as the *law of inertia*. It is usually the first thing about physics which we learn by heart in school, and some of us may remember it:

Every body perseveres in its state of rest, or of uniform motion in a right line, unless it is compelled to change that state by forces impressed thereon.

We have seen that this law of inertia cannot be derived directly from experiment, but only by speculative thinking consistent with observation. The idealized experiment can never be actually performed, although it leads to a profound understanding of real experiments.

From the variety of complex motions in the world around us we choose as our first example uniform motion. This is the simplest, because there are no external forces acting. Uniform motion can, however, never be realized; a stone thrown from a tower, a cart pushed along a road can never move absolutely uniformly because we cannot eliminate the influence of external forces.

In a good mystery story the most obvious clews often lead to the wrong suspects. In our attempts to understand the laws of nature we find, similarly, that the most obvious intuitive explanation is often the wrong one.

Human thought creates an ever-changing picture of the universe. Galileo's contribution was to destroy the intuitive view and replace it by a new one. This is the significance of Galileo's discovery.

But a further question concerning motion arises immediately. If the velocity is no indication of the external forces acting on a body, what is? The answer to this fundamental question was found by Galileo and still more concisely by Newton, and forms a further clew in our investigation.

To find the correct answer we must think a little more deeply about the cart on a perfectly smooth road. In our idealized experiment the uniformity of the motion was due to the absence of all external forces. Let us now imagine that the uniformly moving cart is given a push in the direction of the motion. What happens now? Obviously its speed is increased. Just as obviously, a push in the direction opposite to that of the motion would decrease the speed. In the first case the cart is accelerated by the push, in the second case decelerated, or slowed down. A conclusion follows at once: the action of an external force changes the velocity. Thus not the velocity itself but its change is a consequence of pushing or pulling. Such a force either increases or decreases the velocity according to whether it acts in the direction of motion or in the opposite direction. Galileo saw this clearly and wrote in his *Two New Sciences*:

. . . any velocity once imparted to a moving body will be rigidly maintained as long as the external causes of acceleration or retardation are removed, a condition which is found only on horizontal planes; for in the case of planes which slope downwards there is already present a cause of acceleration; while on planes sloping upward there is retardation; from this it follows that motion along a horizontal plane is perpetual; for, if the velocity be uniform, it cannot be diminished or slackened, much less destroyed.

By following the right clew we achieve a deeper understanding of the problem of motion. The connection between force and the change of velocity and not, as we should think according to our intuition, the connection between force and the velocity itself is the basis of classical mechanics as formulated by Newton.

We have been making use of two concepts which play principal roles in classical mechanics: force and change of velocity. In the further development of science both of these concepts are extended and generalized. They must, therefore, be examined more closely.

What is force? Intuitively, we feel what is meant by this term. The concept arose from the effort of pushing, throwing or pulling; from the muscular sensation accompanying each of these acts. But its generalization goes far beyond these simple examples. We can think of force even without picturing a horse pulling a carriage! We speak of the force of attraction between the sun and the earth, the earth and the moon, and of those forces which cause the tides. We speak of the force by which the earth compels ourselves and all the objects about us to remain within its sphere of influence, and of the force with which the wind makes waves on the sea, or moves the leaves of trees. When and where we observe a change in velocity, an external force, in the general sense, must be held responsible. Newton wrote in his *Principia*:

An impressed force is an action exerted upon a body, in order to change its state, either of rest, or of moving uniformly forward in a right line.

This force consists in the action only; and remains no longer in the body, when the action is over. For a body maintains every new state it acquires, by its *vis inertiae* only. Impressed forces are of different origins; as from percussion, from pressure, from centripetal force.

If a stone is dropped from the top of a tower its motion is by no means uniform; the velocity increases as the stone falls. We conclude: an external force is acting in the direction of the motion. Or, in other words: the earth attracts the stone. Let us take another example. What happens when a stone is thrown straight upward? The velocity decreases until the stone reaches its highest point and begins to fall. This decrease in velocity is caused by the same force as the acceleration of a falling body. In one case the force acts in the direction of the motion, in the other case in the opposite direction. The force is the same, but it causes acceleration or deceleration according to whether the stone is dropped or thrown upward.

VECTORS

All motions we have been considering are *rectilinear*, that is, along a straight line. Now we must go one step further. We gain an understanding of the laws of nature by analyzing the simplest cases and by leaving out of our first attempts all intricate complications. A straight line is simpler than a curve. It is, however, impossible to be satisfied with an understanding of rectilinear motion alone. The motions of the moon, the earth and the planets, just those to which the principles of mechanics have been applied with such brilliant success, are motions along curved paths. Passing from rectilinear motion to motion along a curved path brings new difficulties. We must have the courage to overcome them if we wish to understand the principles of classical mechanics which gave us the first clews and so formed the starting point for the development of science.

Let us consider another idealized experiment, in which a perfect sphere rolls uniformly on a smooth table. We know that if the sphere is given a push, that is, if an external force is applied, the velocity will be changed. Now suppose that the direction of the blow is not, as in the example of the cart, in the line of motion, but in a quite different direction, say, perpendicular to that line. What happens to the sphere? Three stages of the motion can be distinguished: the initial motion, the action of the force, and the final motion after the force has ceased to act. According to the law of inertia the velocities before and after the action of the force are both perfectly uniform. But there is a difference between the uniform motion before and after the action of the force: the direction is changed. The initial path of the sphere and the direction of the force are perpendicular to each other. The final motion will be along neither of these two lines, but somewhere between them, nearer the direction of the force if the blow is a hard one and the initial velocity small, nearer the original line of motion if the blow is gentle and the initial velocity great. Our new conclusion, based on the law of inertia, is: in general the action of an external force changes not only the speed but also the direction of the motion. An understanding of this fact prepares us for the generalization introduced into physics by the concept of *vectors*.

We can continue to use our straightforward method of reasoning. The starting point is again Galileo's law of inertia. We are still far from exhausting the consequences of this valuable clew to the puzzle of motion.

Let us consider two spheres moving in different directions on a smooth table. So as to have a definite picture we may assume the two directions perpendicular to each other. Since there are no external forces acting, the motions are perfectly uniform. Suppose, further, that the speeds are equal, that is, both cover the same distance in the same interval of time. But is it correct to say that the two spheres have the same velocity? The answer can be yes or no! If the speedometers of two cars both show forty miles per hour it is usual to say that they have the same speed or velocity, no matter in which direction they are traveling. But science must create its own language, its own concepts, for its own use. Scientific concepts often begin with those used in ordinary language for the affairs of everyday life, but they develop quite differently. They are transformed and lose the ambiguity associated with them in ordinary language, gaining in rigorousness so that they may be applied to scientific thought.

From the physicist's point of view it is advantageous to say that the velocities of the two spheres moving in different directions are different. Although purely a matter of convention, it is more convenient to say that four cars traveling away from the same traffic circle on different roads do not have the same velocity even though the speeds, as registered on the speedometers, are all forty miles per hour. This differentiation between speed and velocity illustrates how physics, starting with a concept used in everyday life, changes it in a way which proves fruitful in the further development of science.

If a length is measured, the result is expressed as a number of units. The length of a stick may be 3 ft. 7 in.; the weight of some object 2 lb. 3 oz.; a measured time interval so many minutes or seconds. In each of these cases the result of the measurement is expressed by a number. A number alone is, however, insufficient for describing some physical concepts. The recognition of this fact marked a distinct advance in scientific investigation. A direction as well as a number is essential for the characterization of a velocity, for example. Such a quantity, possessing both magnitude

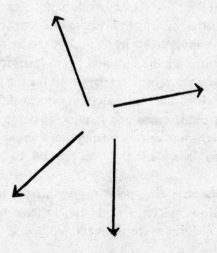

and direction, is called a *vector*. A suitable symbol for
it is an arrow. Velocity may be represented by an
arrow or, briefly speaking, by a vector whose length
in some chosen scale of units is a measure of the speed,
and whose direction is that of the motion.

If four cars diverge with equal speed from a traffic
circle, their velocities can be represented by four vectors of the same length, as seen from our last drawing.
In the scale used one inch stands for 40 m.p.h. In this
way any velocity may be denoted by a vector, and
conversely, if the scale is known, one may ascertain the
velocity from such a vector diagram.

If two cars pass each other on the highway and their
speedometers both show 40 m.p.h. we characterize
their velocities by two different vectors with arrows
pointing in opposite directions. So also the arrows indicating "uptown" and "downtown" subway trains

must point in opposite directions. But all trains moving uptown at different stations or on different avenues with the same speed have the same velocity, which
may be represented by a single vector. There is nothing about a vector to indicate which stations the train
passes or on which of the many parallel tracks it is
running. In other words, according to the accepted
convention, all such vectors, as drawn below, may be
regarded as equal; they lie along the same or parallel
lines, are of equal length, and finally, have arrows

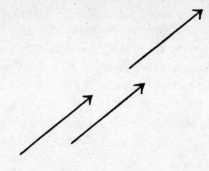

pointing in the same direction. The next figure shows
vectors all different, because they differ either in length

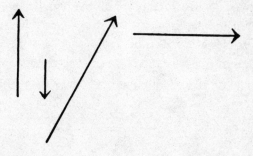

or direction, or both. The same four vectors may be
drawn in another way, in which they all diverge from

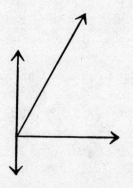

a common point. Since the starting point does not matter, these vectors can represent the velocities of four cars moving away from the same traffic circle, or the velocities of four cars in different parts of the country traveling with the indicated speeds in the indicated directions.

This vector representation may now be used to describe the facts previously discussed concerning rectilinear motion. We talked of a cart, moving uniformly in a straight line and receiving a push in the direction of its motion which increases its velocity. Graphically this may be represented by two vectors, a shorter one denoting the velocity before the push and a longer one in the same direction denoting the velocity after the

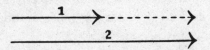

push. The meaning of the dotted vector is clear; it represents the change in velocity for which, as we know, the push is responsible. For the case where the force is directed against the motion, where the motion is slowed down, the diagram is somewhat different.

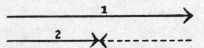

Again the dotted vector corresponds to a change in velocity, but in this case its direction is different. It is clear that not only velocities themselves but also their changes are vectors. But every change in velocity is due to the action of an external force; thus the force must also be represented by a vector. In order to characterize a force it is not sufficient to state how hard we push the cart; we must also say in which direction we push. The force, like the velocity or its change, must be represented by a vector and not by a number alone. Therefore: the external force is also a vector, and must have the same direction as the change in velocity. In the last two drawings the dotted vectors show the direction of the force as truly as they indicate the change in velocity.

Here the skeptic may remark that he sees no advantage in the introduction of vectors. All that has been accomplished is the translation of previously recognized facts into an unfamiliar and complicated language. At this stage it would indeed be difficult to convince him that he is wrong. For the moment he is, in fact, right. But we shall see that just this strange language leads to an important generalization in which vectors appear to be essential.

THE RIDDLE OF MOTION

So long as we deal only with motion along a straight line we are far from understanding the motions observed in nature. We must consider motions along curved paths, and our next step is to determine the laws governing such motions. This is no easy task. In the case of rectilinear motion our concepts of velocity, change of velocity, and force proved most useful. But we do not immediately see how we can apply them to motion along a curved path. It is indeed possible to imagine that the old concepts are unsuited to the description of general motion, and that new ones must be created. Should we try to follow our old path, or seek a new one?

The generalization of a concept is a process very often used in science. A method of generalization is not uniquely determined, for there are usually numerous ways of carrying it out. One requirement, however, must be rigorously satisfied: any generalized concept must reduce to the original one when the original conditions are fulfilled.

We can best explain this by the example with which we are now dealing. We can try to generalize the old concepts of velocity, change of velocity and force for the case of motion along a curved path. Technically, when speaking of curves, we include straight lines. The straight line is a special and trivial example of a curve. If, therefore, velocity, change in velocity and force are introduced for motion along a curved line, then they are automatically introduced for motion along a straight line. But this result must not contradict those results previously obtained. If the curve becomes a straight line, all the generalized concepts must reduce to the familiar ones describing rectilinear motion. But this restriction is not sufficient to determine the generalization uniquely. It leaves open many possibilities. The history of science shows that the simplest generalizations sometimes prove successful and sometimes not. We must first make a guess. In our case it is a simple matter to guess the right method of generalization. The new concepts prove very successful and help us to understand the motion of a thrown stone as well as that of the planets.

And now just what do the words velocity, change in velocity, and force mean in the general case of motion along a curved line? Let us begin with velocity. Along the curve a very small body is moving from left to right. Such a small body is often called a *particle*. The dot on the curve in our drawing shows the position of the particle at some instant of time. What is the veloc-

ity corresponding to this time and position? Again Galileo's clew hints at a way of introducing the velocity. We must, once more, use our imagination and think about an idealized experiment. The particle moves along the curve, from left to right, under the influence of external forces. Imagine that at a given time and at the point indicated by the dot, all these forces suddenly cease to act. Then, the motion must, according to the law of inertia, be uniform. In practice we can, of course, never completely free a body from all external influences. We can only surmise "what would happen if . . . ?" and judge the pertinence of our guess by the conclusions which can be drawn from it and by their agreement with experiment.

The vector in the next drawing indicates the guessed direction of the uniform motion if all external forces

were to vanish. It is the direction of the so-called tangent. Looking at a moving particle through a microscope one sees a very small part of the curve, which appears as a small segment. The tangent is its prolongation. Thus the vector drawn represents the velocity at a given instant. The velocity vector lies on the tangent. Its length represents the magnitude of the velocity, or the speed as indicated, for instance, by the speedometer of a car.

Our idealized experiment about destroying the motion in order to find the velocity vector must not be taken too seriously. It only helps us to understand what we should call the velocity vector and enables us to determine it for a given instant at a given point.

In the next drawing, the velocity vectors for three different positions of a particle moving along a curve

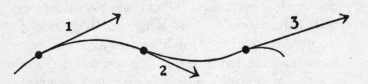

are shown. In this case not only the direction but the magnitude of the velocity, as indicated by the length of the vector, varies during the motion.

Does this new concept of velocity satisfy the requirement formulated for all generalizations? That is: does it reduce to the familiar concept if the curve becomes a straight line? Obviously it does. The tangent to a straight line is the line itself. The velocity vector lies in the line of the motion, just as in the case of the moving cart or the rolling spheres.

The next step is the introduction of the change in velocity of a particle moving along a curve. This also may be done in various ways, from which we choose the simplest and most convenient. The last drawing

showed several velocity vectors representing the motion at various points along the path. The first two of
these may be drawn again so that they have a common

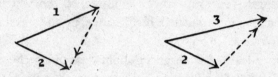

starting point, as we have seen is possible with vectors.
The dotted vector, we call the change in velocity. Its
starting point is the end of the first vector and its end
point the end of the second vector. This definition of
the change in velocity may, at first sight, seem artificial
and meaningless. It becomes much clearer in the special
case in which vectors (1) and (2) have the same direction. This, of course, means going over to the case of
straight-line motion. If both vectors have the same
initial point the dotted vector again connects their end
points. The drawing is now identical with that on page

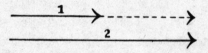

17, and the previous concept is regained as a special case
of the new one. We may remark that we had to separate the two lines in our drawing since otherwise they
would coincide and be indistinguishable.

We now have to take the last step in our process of
generalization. It is the most important of all the
guesses we have had to make so far. The connection
between force and change in velocity has to be established so that we can formulate the clew which will

enable us to understand the general problem of motion.

The clew to an explanation of motion along a straight line was simple: external force is responsible for change in velocity; the force vector has the same direction as the change. And now what is to be regarded as the clew to curvilinear motion? Exactly the same! The only difference is that change of velocity has now a broader meaning than before. A glance at the dotted vectors of the last two drawings shows this point clearly. If the velocity is known for all points along the curve, the direction of the force at any point can be deduced at once. One must draw the velocity vectors for two instants separated by a very short time-interval and therefore corresponding to positions very near each other. The vector from the end point of the first to that of the second indicates the direction of the acting force. But it is essential that the two velocity vectors should be separated only by a "very short" time-interval. The rigorous analysis of such words as "very near," "very short" is far from simple. Indeed it was this analysis which led Newton and Leibnitz to the discovery of differential calculus.

It is a tedious and elaborate path which leads to the generalization of Galileo's clew. We cannot show here how abundant and fruitful the consequences of this generalization have proved. Its application leads to simple and convincing explanations of many facts previously incoherent and misunderstood.

From the extremely rich variety of motions we shall take only the simplest and apply to their explanation the law just formulated.

A bullet shot from a gun, a stone thrown at an angle,

a stream of water emerging from a hose, all describe familiar paths of the same type, the parabola. Imagine a speedometer attached to a stone, for example, so that its velocity vector may be drawn for any instant.

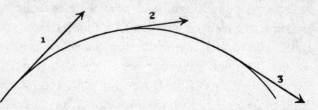

The result may well be that represented in the last drawing. The direction of the force acting on the stone is just that of the change in velocity, and we have seen how it may be determined. The result, shown in the next drawing, indicates that the force is vertical,

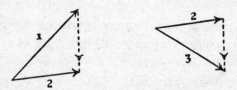

and directed downward. It is exactly the same as when a stone is allowed to fall from the top of a tower. The paths are quite different, as also are the velocities, but the change in velocity has the same direction, that is, toward the center of the earth.

A stone attached to the end of a string and swung around in a horizontal plane moves in a circular path.

All the vectors in the diagram representing this motion have the same length if the speed is uniform. Nevertheless, the velocity is not uniform, for the path is not a straight line. Only in uniform, rectilinear motion are there no forces involved. Here, however, there

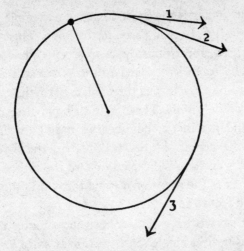

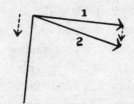

are, and the velocity changes not in magnitude but in direction. According to the law of motion there must be some force responsible for this change, a force in this case between the stone and the hand holding the string. A further question arises immediately: in what direction does the force act? Again a vector diagram shows the answer. The velocity vectors for two very near points are drawn, and the change of velocity

found. This last vector is seen to be directed along the string toward the center of the circle, and is always perpendicular to the velocity vector, or tangent. In other words the hand exerts a force on the stone by means of the string.

Very similar is the more important example of the revolution of the moon around the earth. This may be represented approximately as uniform circular motion. The force is directed toward the earth for the same reason that it was directed toward the hand in our former example. There is no string connecting the earth and the moon, but we can imagine a line between the centers of the two bodies; the force lies along this line and is directed toward the center of the earth, just as the force on a stone thrown in the air or dropped from a tower.

All that we have said concerning motion can be summed up in a single sentence. *Force and change of velocity are vectors having the same direction.* This is the initial clew to the problem of motion, but it certainly does not suffice for a thorough explanation of all motions observed. The transition from Aristotle's line of thought to that of Galileo formed a most important cornerstone in the foundation of science. Once this break was made, the line of further development was clear. Our interest here lies in the first stages of development, in following initial clews, in showing how new physical concepts are born in the painful struggle with old ideas. We are concerned only with pioneer work in science, which consists of finding new and unexpected paths of development; with the adventures in scientific thought which create an ever-changing picture of the universe. The initial and fundamental steps are always of a revolutionary character. Scientific imagination finds old concepts too confining, and replaces them by new ones. The continued development along any line already initiated is more in the nature of evolution, until the next turning point is reached when a still newer field must be conquered. In order to understand, however, what reasons and what difficulties force a change in important concepts, we must know not only the initial clews, but also the conclusions which can be drawn.

One of the most important characteristics of modern physics is that the conclusions drawn from initial clews are not only qualitative but also quantitative. Let us again consider a stone dropped from a tower. We have seen that its velocity increases as it falls, but we should like to know much more. Just how great is this change? And what is the position and the velocity of the stone at any time after it begins to fall? We wish to be able to predict events and to determine by experiment whether observation confirms these predictions and thus the initial assumptions.

To draw quantitative conclusions we must use the language of mathematics. Most of the fundamental ideas of science are essentially simple, and may, as a rule, be expressed in a language comprehensible to everyone. To follow up these ideas demands the knowledge of a highly refined technique of investigation. Mathematics as a tool of reasoning is necessary if we wish to draw conclusions which may be compared with experiment. So long as we are concerned only with fundamental physical ideas we may avoid the language of mathematics. Since in these pages we do this consistently, we must occasionally restrict ourselves to quoting, without proof, some of the results necessary for an understanding of important clews arising in the further development. The price which must be paid for abandoning the language of mathematics is a loss in precision, and the necessity of sometimes quoting results without showing how they were reached.

A very important example of motion is that of the earth around the sun. It is known that the path is a closed curve, called the ellipse. The construction of a vector diagram of the change in velocity shows that the force on the earth is directed toward the sun. But

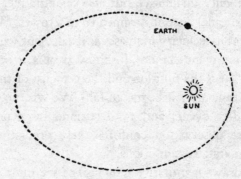

this, after all, is scant information. We should like to be able to predict the position of the earth and the other planets for any arbitrary instant of time, we should like to predict the date and duration of the next solar eclipse and many other astronomical events. It is possible to do these things, but not on the basis of our initial clew alone, for it is now necessary to know not only the direction of the force but also its absolute value, its magnitude. It was Newton who made the inspired guess on this point. According to his *law of gravitation* the force of attraction between two bodies depends in a simple way on their distance from each other. It becomes smaller when the distance increases. To be specific it becomes $2 \times 2 = 4$ times smaller if the distance is doubled, $3 \times 3 = 9$ times smaller if the distance is made three times as great.

Thus we see that in the case of gravitational force we have succeeded in expressing, in a simple way, the dependence of the force on the distance between the moving bodies. We proceed similarly in all other cases where forces of different kinds, for instance, electric, magnetic, and the like, are acting. We try to use a simple expression for the force. Such an expression is justified only when the conclusions drawn from it are confirmed by experiment.

But this knowledge of the gravitational force alone is not sufficient for a description of the motion of the planets. We have seen that vectors representing force and change in velocity for any short interval of time have the same direction, but we must follow Newton one step further and assume a simple relation between their lengths. Given all other conditions the same, that is, the same moving body and changes considered over equal time intervals, then, according to Newton, the change of velocity is proportional to the force.

Thus just two complementary guesses are needed for quantitative conclusions concerning the motion of the planets. One is of a general character, stating the connection between force and change in velocity. The other is special, and states the exact dependence of the particular kind of force involved on the distance between the bodies. The first is Newton's general law of motion, the second his law of gravitation. Together they determine the motion. This can be made clear by the following somewhat clumsy-sounding reasoning. Suppose that at a given time the position and velocity of a planet can be determined, and that the force is known. Then, according to Newton's laws we know the change in velocity during a short time interval.

Knowing the initial velocity and its change, we can find the velocity and position of the planet at the end of the time interval. By a continued repetition of this process the whole path of the motion may be traced without further recourse to observational data. This is, in principle, the way mechanics predicts the course of a body in motion, but the method used here is hardly practical. In practice such a step-by-step procedure would be extremely tedious as well as inaccurate. Fortunately, it is quite unnecessary; mathematics furnishes a short cut, and makes possible precise description of the motion in much less ink than we use for a single sentence. The conclusions reached in this way can be proved or disproved by observation.

The same kind of external force is recognized in the motion of a stone falling through the air and in the revolution of the moon in its orbit, namely, that of the earth's attraction for material bodies. Newton recognized that the motions of falling stones, of the moon, and of planets are only very special manifestations of a universal gravitational force acting between any two bodies. In simple cases the motion may be described and predicted by the aid of mathematics. In remote and extremely complicated cases, involving the action of many bodies on each other, a mathematical description is not so simple, but the fundamental principles are the same.

We find the conclusions, at which we arrived by following our initial clews, realized in the motion of a thrown stone, in the motion of the moon, the earth, and the planets.

It is really our whole system of guesses which is to be either proved or disproved by experiment. No one of the assumptions can be isolated for separate testing. In the case of the planets moving around the sun it is found that the system of mechanics works splendidly. Nevertheless we can well imagine that another system, based on different assumptions, might work just as well.

Physical concepts are free creations of the human mind, and are not, however it may seem, uniquely determined by the external world. In our endeavor to understand reality we are somewhat like a man trying to understand the mechanism of a closed watch. He sees the face and the moving hands, even hears its ticking, but he has no way of opening the case. If he is ingenious he may form some picture of a mechanism which could be responsible for all the things he observes, but he may never be quite sure his picture is the only one which could explain his observations. He will never be able to compare his picture with the real mechanism and he cannot even imagine the possibility or the meaning of such a comparison. But he certainly believes that, as his knowledge increases, his picture of reality will become simpler and simpler and will explain a wider and wider range of his sensuous impressions. He may also believe in the existence of the ideal limit of knowledge and that it is approached by the human mind. He may call this ideal limit the objective truth.

ONE CLEW REMAINS

When first studying mechanics one has the impression that everything in this branch of science is simple, fundamental and settled for all time. One would hardly suspect the existence of an important clew which no one noticed for three hundred years. The neglected clew is connected with one of the fundamental concepts of mechanics, that of *mass*.

Again we return to the simple idealized experiment of the cart on a perfectly smooth road. If the cart is initially at rest and then given a push, it afterwards moves uniformly with a certain velocity. Suppose that the action of the force can be repeated as many times as desired, the mechanism of pushing acting in the same way and exerting the same force on the same cart. However many times the experiment is repeated the final velocity is always the same. But what happens if the experiment is changed, if previously the cart was empty and now it is loaded? The loaded cart will have a smaller final velocity than the empty one. The conclusion is: if the same force acts on two different bodies, both initially at rest, the resulting velocities will not be the same. We say that the velocity depends on the mass of the body, being smaller if the mass is greater.

We know, therefore, at least in theory, how to determine the mass of a body or, more exactly, how many times greater one mass is than another. We have identical forces acting on two resting masses. Finding that the velocity of the first mass is three times greater than that of the second we conclude that the first mass is three times smaller than the second. This is certainly not a very practical way of determining the ratio of two masses. We can, nevertheless, well imagine having done it in this, or in some similar way, based upon the application of the law of inertia.

How do we really determine mass in practice? Not, of course, in the way just described. Everyone knows the correct answer. We do it by weighing on a scale.

Let us discuss in more detail the two different ways of determining mass.

The first experiment had nothing whatever to do with gravity, the attraction of the earth. The cart moves along a perfectly smooth and horizontal plane after the push. Gravitational force, which causes the cart to stay on the plane, does not change, and plays no role in the determination of the mass. It is quite different with weighing. We could never use a scale if the earth did not attract bodies, if gravity did not exist. The difference between the two determinations of mass is that the first has nothing to do with the force of gravity while the second is based essentially on its existence.

We ask: if we determine the ratio of two masses in both ways described above do we obtain the same result? The answer given by experiment is quite clear. The results are exactly the same! This conclusion could not have been foreseen, and is based on observation, not reason. Let us, for the sake of simplicity, call the mass determined in the first way the *inertial mass* and that determined in the second way the *gravitational mass*. In our world it happens that they are equal but we can well imagine that this should not have been the case at all. Another question arises immediately: is this identity of the two kinds of mass purely accidental, or does it have a deeper significance? The answer, from the point of view of classical physics, is: the identity of the two masses is accidental and no deeper significance should be attached to it. The answer of modern physics is just the opposite: the identity of the two masses is fundamental and forms a new and essential clew leading to a more profound understanding. This was, in fact, one of the most important clews from which the so-called general theory of relativity was developed.

A mystery story seems inferior if it explains strange events as accidents. It is certainly more satisfying to have the story follow a rational pattern. In exactly the same way a theory which offers an explanation for the identity of gravitational and inertial mass is superior to one which interprets their identity as accidental, provided, of course, that the two theories are equally consistent with observed facts.

Since this identity of inertial and gravitational mass was fundamental for the formulation of the theory of relativity we are justified in examining it a little more closely here. What experiments prove convincingly that the two masses are the same? The answer lies in Galileo's old experiment in which he dropped different masses from a tower. He noticed that the time required for the fall was always the same, that the motion of a falling body does not depend on the mass. To link this simple but highly important experimental result with the identity of the two masses needs some rather intricate reasoning.

A body at rest gives way before the action of an external force, moving and attaining a certain velocity. It yields more or less easily, according to its inertial mass, resisting the motion more strongly if the mass is large than if it is small. We may say, without pretending to be rigorous: the readiness with which a body responds to the call of an external force depends on its inertial mass. If it were true that the earth attracts all bodies with the same force, that of greatest inertial mass would move more slowly in falling than any other. But this is not the case: all bodies fall in the same way. This means that the force by which the earth attracts different masses must be different. Now the earth attracts a stone with the force of gravity and knows nothing about its inertial mass. The "calling" force of the earth depends on the gravitational mass. The "answering" motion of the stone depends on the inertial mass. Since the "answering" motion is always the same —all bodies dropped from the same height fall in the same way—it must be deduced that gravitational mass and inertial mass are equal.

More pedantically a physicist formulates the same conclusion: the acceleration of a falling body increases in proportion to its gravitational mass and decreases in proportion to its inertial mass. Since all falling bodies have the same constant acceleration, the two masses must be equal.

In our great mystery story there are no problems wholly solved and settled for all time. After three hundred years we had to return to the initial problem of motion, to revise the procedure of investigation, to find clews which had been overlooked, thereby reaching a different picture of the surrounding universe.

IS HEAT A SUBSTANCE?

Here we begin to follow a new clew, one originating in the realm of heat phenomena. It is impossible, however, to divide science into separate and unrelated sections. Indeed, we shall soon find that the new concepts introduced here are interwoven with those already familiar, and with those we shall still meet. A line of thought developed in one branch of science can very often be applied to the description of events apparently quite different in character. In this process the original concepts are often modified so as to advance the understanding both of those phenomena from which they sprang and of those to which they are newly applied.

The most fundamental concepts in the description of heat phenomena are *temperature* and *heat*. It took an unbelievably long time in the history of science for these two to be distinguished, but once this distinction was made rapid progress resulted. Although these concepts are now familiar to everyone we shall examine them closely, emphasizing the differences between them.

Our sense of touch tells us quite definitely that one body is hot and another cold. But this is a purely qualitative criterion, not sufficient for a quantitative description and sometimes even ambiguous. This is shown by a well-known experiment: we have three vessels containing, respectively, cold, warm and hot water. If we dip one hand into the cold water and the other into the hot, we receive a message from the first that it is cold and from the second that it is hot. If we then dip both hands into the same warm water we receive two contradictory messages, one from each hand. For the same reason an Eskimo and a native of some equatorial country meeting in New York on a spring day would hold different opinions as to whether the climate was hot or cold. We settle all such questions by the use of a thermometer, an instrument designed in a primitive form by Galileo. Here again that familiar name! The use of a thermometer is based on some obvious physical assumptions. We shall recall them by quoting a few lines from lectures given about a hundred and fifty years ago by Black, who contributed a great deal toward clearing up the difficulties connected with the two concepts, heat and temperature:

By the use of this instrument we have learned, that if we take 1000, or more, different kinds of matter, such as metals, stones, salts, woods, feathers, wool, water and a variety of other fluids, although they be all at first of different *heats*, let them be placed together in the same room without a fire, and into which the sun does not shine, the heat will be communicated from the hotter of these bodies to the colder, during some hours perhaps, or the course of a day, at the end of which time, if we apply a thermometer to them all in succession, it will point precisely to the same degree.

The italicized word *heats* should, according to present-day nomenclature, be replaced by the word *temperatures*.

A physician taking the thermometer from a sick man's mouth might reason like this: "The thermometer indicates its own temperature by the length of its column of mercury. We assume that the length of the mercury column increases in proportion to the increase in temperature. But the thermometer was for a few minutes in contact with my patient, so that both patient and thermometer have the same temperature. I conclude, therefore, that my patient's temperature is that registered on the thermometer." The doctor probably acts mechanically, but he applies physical principles without thinking about it.

But does the thermometer contain the same amount of heat as the body of the man? Of course not. To assume that two bodies contain equal quantities of heat just because their temperatures are equal would, as Black remarked, be

taking a very hasty view of the subject. It is confounding the quantity of heat in different bodies with its general strength or intensity, though it is plain that these are two different things, and should always be distinguished, when we are thinking of the distribution of heat.

An understanding of this distinction can be gained by considering a very simple experiment. A pound of water placed over a gas flame takes some time to change from room temperature to the boiling point. A much longer time is required for heating twelve pounds, say, of water in the same vessel by means of the same flame. We interpret this fact as indicating that now more of "something" is needed and we call this "something"—*heat*.

A further important concept, *specific heat*, is gained by the following experiment: let one vessel contain a pound of water and another a pound of mercury, both to be heated in the same way. The mercury gets hot much more quickly than the water, showing that less "heat" is needed to raise the temperature by one degree. In general, different amounts of "heat" are required to change by one degree, say from 40 to 41 degrees Fahrenheit, the temperatures of different substances such as water, mercury, iron, copper, wood, etc., all of the same mass. We say that each substance has its individual *heat capacity*, or *specific heat*.

Once having gained the concept of heat we can investigate its nature more closely. We have two bodies, one hot, the other cold, or more precisely, one of a higher temperature than the other. We bring them into contact and free them from all other external influences. Eventually they will, we know, reach the same temperature. But how does this take place? What happens between the instant they are brought into contact and the achievement of equal temperatures? The picture of heat "flowing" from one body to another suggests itself, like water flowing from a higher level to a lower. This picture, though primitive, seems to fit many of the facts, so that the analogy runs:

Water — Heat
Higher level — Higher temperature
Lower level — Lower temperature

The flow proceeds until both levels, that is, both temperatures, are equal. This naïve view can be made more useful by quantitative considerations. If definite masses of water and alcohol, each at a definite temperature, are mixed together, a knowledge of the specific heats will lead to a prediction of the final temperature of the mixture. Conversely, an observation of the final temperature, together with a little algebra, would enable us to find the ratio of the two specific heats.

We recognize in the concept of heat which appears here a similarity to other physical concepts. Heat is, according to our view, a substance, such as mass in mechanics. Its quantity may change or not, like money put aside in a safe or spent. The amount of money in a safe will remain unchanged so long as the safe remains locked, and so will the amounts of mass and heat in an isolated body. The ideal thermos bottle is analogous to such a safe. Furthermore, just as the mass of an isolated system is unchanged even if a chemical transformation takes place, so heat is conserved even though it flows from one body to another. Even if heat is not used for raising the temperature of a body but for melting ice, say, or changing water into steam, we can still think of it as a substance and regain it entirely by freezing the water or liquefying the steam. The old names, latent heat of melting or vaporization, show that these concepts are drawn from the picture of heat as a substance. Latent heat is temporarily hidden, like money put away in a safe, but available for use if one knows the lock combination.

But heat is certainly not a substance in the same sense as mass. Mass can be detected by means of scales, but what of heat? Does a piece of iron weigh more when red-hot than when ice-cold? Experiment shows that it does not. If heat is a substance at all, it is a weightless one. The "heat-substance" was usually called *caloric* and is our first acquaintance among a whole family of weightless substances. Later we shall have occasion to follow the history of the family, its rise and fall. It is sufficient now to note the birth of this particular member.

The purpose of any physical theory is to explain as wide a range of phenomena as possible. It is justified in so far as it does make events understandable. We have seen that the substance theory explains many of the heat phenomena. It will soon become apparent, however, that this again is a false clew, that heat cannot be regarded as a substance, even weightless. This is clear if we think about some simple experiments which marked the beginning of civilization.

We think of a substance as something which can be

neither created nor destroyed. Yet primitive man created by friction sufficient heat to ignite wood. Examples of heating by friction are, as a matter of fact, much too numerous and familiar to need recounting. In all these cases some quantity of heat is created, a fact difficult to account for by the substance theory. It is true that a supporter of this theory could invent arguments to account for it. His reasoning would run something like this: "The substance theory can explain the apparent creation of heat. Take the simplest example of two pieces of wood rubbed one against the other. Now rubbing is something which influences the wood and changes its properties. It is very likely that the properties are so modified that an unchanged quantity of heat comes to produce a higher temperature than before. After all, the only thing we notice is the rise in temperature. It is possible that the friction changes the specific heat of the wood and not the total amount of heat."

At this stage of the discussion it would be useless to argue with a supporter of the substance theory, for this is a matter which can be settled only by experiment. Imagine two identical pieces of wood and suppose equal changes of temperature are induced by different methods; in one case by friction and in the other by contact with a radiator, for example. If the two pieces have the same specific heat at the new temperature the whole substance theory must break down. There are very simple methods for determining specific heats, and the fate of the theory depends on the result of just such measurements. Tests which are capable of pronouncing a verdict of life or death on a theory occur frequently in the history of physics, and are called *crucial* experiments. The crucial value of an experiment is revealed only by the way the question is formulated, and only one theory of the phenomena can be put on trial by it. The determination of the specific heats of two bodies of the same kind, at equal temperatures attained by friction and heat flow respectively, is a typical example of a crucial experiment. This experiment was performed about a hundred and fifty years ago by Rumford, and dealt a death blow to the substance theory of heat.

An extract from Rumford's own account tells the story:

It frequently happens, that in the ordinary affairs and occupations of life, opportunities present themselves of contemplating some of the most curious operations of Nature; and very interesting philosophical experiments might often be made, almost without trouble or expense, by means of machinery contrived for the mere mechanical purposes of the arts and manufactures.

I have frequently had occasion to make this observation; and am persuaded, that a habit of keeping the eyes open to every thing that is going on in the ordinary course of the business of life has oftener led, as it were by accident, or in the playful excursions of the imagination, put into action by contemplating the most common appearances, to useful doubts, and sensible schemes for investigation and improvement, than all the more intense meditations of philosophers, in the hours expressly set apart for study. . . .

Being engaged, lately, in superintending the boring of cannon, in the workshops of the military arsenal at Munich, I was struck with the very considerable degree of Heat which a brass gun acquires, in a short time, in being bored; and with the still more intense Heat (much greater than that of boiling water, as I found by experiment) of the metallic chips separated from it by the borer. . . .

From whence comes the Heat actually produced in the mechanical operation above mentioned?

Is it furnished by the metallic chips which are separated by the borer from the solid mass of metal?

If this were the case, then, according to the modern doctrines of latent Heat, and of caloric, the capacity ought not only to be changed, but the change undergone by them should be sufficiently great to account for all the Heat produced.

But no such change had taken place; for I found, upon taking equal quantities, by weight, of these chips, and of thin slips of the same block of metal separated by means of a fine saw and putting them, at the same temperature (that of boiling water), into equal quantities of cold water (that is to say, at the temperature of 59½° F.) the portion of water into which the chips were put was not, to all appearance, heated either less or more than the other portion, in which the slips of metal were put.

Finally we reach his conclusion:

And, in reasoning on this subject, we must not forget to consider that most remarkable circumstance, that the source of the Heat generated by friction, in these Experiments, appeared evidently to be *inexhaustible.*

It is hardly necessary to add, that anything which any *insulated* body, or system of bodies, can continue to furnish *without limitation,* cannot possibly be a *material substance;* and it appears to me to be extremely difficult, if not quite impossible, to form any distinct idea of anything, capable of being excited and communicated, in the manner the Heat was excited and communicated in these Experiments, except it be MOTION.

Thus we see the breakdown of the old theory, or to be more exact, we see that the substance theory is limited to problems of heat flow. Again, as Rumford has intimated, we must seek a new clew. To do this, let us leave for the moment the problem of heat and return to mechanics.

THE ROLLER-COASTER

Let us trace the motion of that popular thrill-giver, the roller-coaster. A small car is lifted or driven to the highest point of the track. When set free it starts rolling down under the force of gravity, and then goes up and down along a fantastically curved line, giving the occupants a thrill by the sudden changes in velocity. Every roller-coaster has its highest point, that from which it starts. Never again, throughout the whole course of the motion will it reach the same height. A complete description of the motion would be very complicated. On the one hand is the mechanical side of the problem, the changes of velocity and position in time. On the other there is friction and therefore the creation of heat, on the rail and in the wheels. The only significant reason for dividing the physical process into these two aspects is to make possible the use of the concepts previously discussed. The division leads to an idealized experiment, for a physical process in which only the mechanical aspect appears can be only imagined but never realized.

For the idealized experiment we may imagine that someone has learned to eliminate entirely the friction which always accompanies motion. He decides to apply his discovery to the construction of a roller-coaster, and must find out for himself how to build one. The car is to run up and down, with its starting point, say, at one hundred feet above ground level. He soon discovers by trial and error that he must follow a very simple rule: he may build his track in whatever path he pleases so long as no point is higher than the starting point. If the car is to proceed freely to the

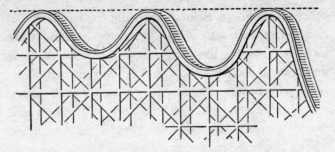

end of the course its height may attain a hundred feet as many times as he likes, but never exceed it. The initial height can never be reached by a car on an actual track because of friction, but our hypothetical engineer need not consider that.

Let us follow the motion of the idealized car on the idealized roller-coaster as it begins to roll downward from the starting point. As it moves its distance from the ground diminishes, but its speed increases. This sentence at first sight may remind us of one from a language lesson: "I have no pencil, but you have six oranges." It is not so stupid, however. There is no connection between my having no pencil and your having six oranges, but there is a very real correlation between the distance of the car from the ground and its speed. We can calculate the speed of the car at any moment if we know how high it happens to be above the ground, but we omit this point here because of its quantitative character which can best be expressed by mathematical formulae.

At its highest point the car has zero velocity and is one hundred feet from the ground. At the lowest possible point it is no distance from the ground, and has its greatest velocity. These facts may be expressed in other terms. At its highest point the car has *potential energy* but no *kinetic energy* or energy of motion. At its lowest point it has the greatest kinetic energy and no potential energy whatever. At all intermediate positions, where there is some velocity and some elevation, it has both kinetic and potential energy. The potential energy increases with the elevation, while the kinetic energy becomes greater as the velocity increases. The principles of mechanics suffice to explain the motion. Two expressions for energy occur in the mathematical description, each of which changes, although the sum does not vary. It is thus possible to introduce mathematically and rigorously the concepts of potential energy, depending on position, and kinetic energy, depending on velocity. The introduction of the two names is, of course, arbitrary and justified only by convenience. The sum of the two quantities remains unchanged, and is called a constant of the motion. The total energy, kinetic plus potential, is like a substance, for example, money kept intact as to amount but changed continually from one currency to another, say from dollars to pounds and back again, according to a well-defined rate of exchange.

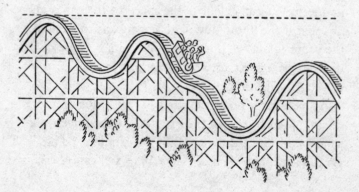

In the real roller-coaster, where friction prevents the car from again reaching as high a point as that from which it started, there is still a continuous change between kinetic and potential energy. Here, however, the sum does not remain constant, but grows smaller. Now one important and courageous step more is needed to relate the mechanical and heat aspects of motion. The wealth of consequences and generalizations from this step will be seen later.

Something more than kinetic and potential energies is now involved, namely, the heat created by friction. Does this heat correspond to the diminution in mechanical energy, that is kinetic and potential energy? A new guess is imminent. If heat may be regarded as a form of energy, perhaps the sum of all three, heat, kinetic and potential energies, remains constant. Not heat alone, but heat and other forms of energy taken together are, like a substance, indestructible. It is as if a man must pay himself a commission in francs for changing dollars to pounds, the commission money also being saved so that the sum of dollars, pounds, and francs is a fixed amount according to some definite exchange rate.

The progress of science has destroyed the older concept of heat as a substance. We try to create a new substance, energy, with heat as one of its forms.

THE RATE OF EXCHANGE

Less than a hundred years ago the new clew which led to the concept of heat as a form of energy was guessed by Mayer and confirmed experimentally by Joule. It is a strange coincidence that nearly all the fundamental work concerned with the nature of heat was done by non-professional physicists who regarded physics merely as their great hobby. There was the versatile Scotsman Black, the German physician Mayer, and the great American adventurer Count Rumford, who afterward lived in Europe, and among other activities, became Minister of War for Bavaria. There was also the English brewer Joule who, in his spare time, performed some most important experiments concerning the conservation of energy.

Joule verified by experiment the guess that heat is a form of energy, and determined the rate of exchange. It is worth our while to see just what his results were.

The kinetic and potential energy of a system together constitute its *mechanical* energy. In the case of the roller-coaster we made a guess that some of the mechanical energy was converted into heat. If this is right there must be here and in all other similar physical processes a definite *rate of exchange* between the two. This is rigorously a quantitative question, but the fact that a given quantity of mechanical energy can be changed into a definite amount of heat is highly important. We should like to know what number expresses the rate of exchange, i.e. how much heat we obtain from a given amount of mechanical energy.

The determination of this number was the object of Joule's researches. The mechanism of one of his experiments is very much like that of a weight clock. The winding of such a clock consists of elevating two weights, thereby adding potential energy to the system. If the clock is not further interfered with it may be regarded as a closed system. Gradually the weights fall and the clock runs. At the end of a certain time the weights will have reached their lowest position and the clock will have stopped. What has happened to the energy? The potential energy of the weights has changed into kinetic energy of the mechanism, and has then gradually been dissipated as heat.

A clever alteration in this sort of mechanism enabled Joule to measure the heat lost and thus the rate of exchange. In his apparatus two weights caused a paddle wheel to turn while immersed in water. The potential energy of the weights was changed into kinetic energy of the movable parts, and thence into heat

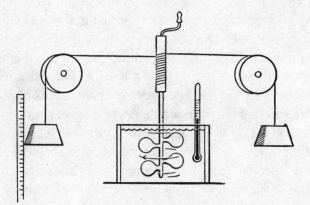

which raised the temperature of the water. Joule measured this change of temperature and, making use of the known specific heat of water, calculated the amount of heat absorbed. He summarized the results of many trials as follows:

1st. That the quantity of heat produced by the friction of bodies, whether solid or liquid, is always proportional to the quantity of force [by force Joule means energy] expended. And

2nd. That the quantity of heat capable of increasing the temperature of a pound of water (weighed in vacuo and taken at between 55° and 60°) by 1° Fahr. requires for its evolution the expenditure of a mechanical force [energy] represented by the fall of 772 lb. through the space of one foot.

In other words the potential energy of 772 pounds elevated one foot above the ground is equivalent to the quantity of heat necessary to raise the temperature of one pound of water from 55° F. to 56° F. Later experimenters were capable of somewhat greater accuracy, but the mechanical equivalent of heat is essentially what Joule found in his pioneer work.

Once this important work was done, further progress was rapid. It was soon recognized that these kinds of energy, mechanical and heat, are only two of its many forms. Everything which can be converted into either of them is also a form of energy. The radiation given off by the sun is energy, for part of it is transformed into heat on the earth. An electric current possesses energy, for it heats a wire or turns the wheels of a motor. Coal represents chemical energy, liberated as heat when the coal burns. In every event in nature one form of energy is being converted into another, always at some well-defined rate of exchange. In a closed system, one isolated from external influences, the energy is conserved and thus behaves like a substance. The sum of all possible forms of energy in such a system is constant, although the amount of any one kind may be changing. If we regard the whole universe as a closed system we can proudly announce with the physicists of the nineteenth century that the energy of the universe is invariant, that no part of it can ever be created or destroyed.

Our two concepts of substance are, then, *matter* and *energy*. Both obey conservation laws: An isolated system cannot change either in mass or in total energy. Matter has weight but energy is weightless. We have therefore two different concepts and two conservation laws. Are these ideas still to be taken seriously? Or has this apparently well-founded picture been changed in the light of newer developments? It has! Further changes in the two concepts are connected with the theory of relativity. We shall return to this point later.

THE PHILOSOPHICAL BACKGROUND

The results of scientific research very often force a change in the philosophical view of problems which extend far beyond the restricted domain of science itself. What is the aim of science? What is demanded of a theory which attempts to describe nature? These questions, although exceeding the bounds of physics, are intimately related to it, since science forms the material from which they arise. Philosophical generalizations must be founded on scientific results. Once formed and widely accepted, however, they very often influence the further development of scientific thought by indicating one of the many possible lines of procedure. Successful revolt against the accepted view results in unexpected and completely different developments, becoming a source of new philosophical aspects. These remarks necessarily sound vague and pointless until illustrated by examples quoted from the history of physics.

We shall here try to describe the first philosophical ideas on the aim of science. These ideas greatly influenced the development of physics until nearly a hundred years ago, when their discarding was forced by new evidence, new facts and theories, which in their turn formed a new background for science.

In the whole history of science from Greek philosophy to modern physics there have been constant attempts to reduce the apparent complexity of natural phenomena to some simple fundamental ideas and relations. This is the underlying principle of all natural philosophy. It is expressed even in the work of the Atomists. Twenty-three centuries ago Democritus wrote:

By convention sweet is sweet, by convention bitter is bitter, by convention hot is hot, by convention cold is cold, by convention color is color. But in reality there are atoms and the void. That is, the objects of sense are supposed to be real and it is customary to regard them as such, but in truth they are not. Only the atoms and the void are real.

This idea remains in ancient philosophy nothing more than an ingenious figment of the imagination. Laws of nature relating subsequent events were unknown to the Greeks. Science connecting theory and experiment really began with the work of Galileo. We have followed the initial clews leading to the laws of motion. Throughout two hundred years of scientific research force and matter were the underlying concepts in all endeavors to understand nature. It is impossible to imagine one without the other because matter demonstrates its existence as a source of force by its action on other matter.

Let us consider the simplest example: two particles

with forces acting between them. The easiest forces to imagine are those of attraction and repulsion. In both cases the force vectors lie on a line connecting the material points. The demand for simplicity leads to the picture of particles attracting or repelling each other;

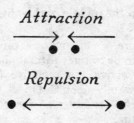

Attraction

Repulsion

any other assumption about the direction of the acting forces would give a much more complicated picture. Can we make an equally simple assumption about the length of the force vectors? Even if we want to avoid too special assumptions we can still say one thing: The force between any two given particles depends only on the distance between them, like gravitational forces. This seems simple enough. Much more complicated forces could be imagined, such as those which might depend not only on the distance but also on the velocities of the two particles. With matter and force as our fundamental concepts we can hardly imagine simpler assumptions than that forces act along the line connecting the particles and depend only on the distance. But is it possible to describe all physical phenomena by forces of this kind alone?

The great achievements of mechanics in all its branches, its striking success in the development of astronomy, the application of its ideas to problems apparently different and non-mechanical in character, all these things contributed to the belief that it *is* possible to describe all natural phenomena in terms of simple forces between unalterable objects. Throughout the two centuries following Galileo's time such an endeavor, conscious or unconscious, is apparent in nearly all scientific creation. This was clearly formulated by Helmholtz about the middle of the nineteenth century:

Finally, therefore, we discover the problem of physical material science to be to refer natural phenomena back to unchangeable attractive and repulsive forces whose intensity depends wholly upon distance. The solubility of this problem is the condition of the complete comprehensibility of nature.

Thus, according to Helmholtz, the line of development of science is determined and follows strictly a fixed course:

And its vocation will be ended as soon as the reduction of natural phenomena to simple forces is complete and the proof given that this is the only reduction of which the phenomena are capable.

This view appears dull and naïve to a twentieth-century physicist. It would frighten him to think that the great adventure of research could be so soon finished, and an unexciting if infallible picture of the universe established for all time.

Although these tenets would reduce the description of all events to simple forces, they do leave open the question of just how the forces should depend on distance. It is possible that for different phenomena this dependence is different. The necessity of introducing many different kinds of force for different events is certainly unsatisfactory from a philosophical point of view. Nevertheless this so-called *mechanical view*, most clearly formulated by Helmholtz, played an important role in its time. The development of the kinetic theory of matter is one of the greatest achievements directly influenced by the mechanical view.

Before witnessing its decline, let us provisionally accept the point of view held by the physicists of the past century and see what conclusions we can draw from their picture of the external world.

THE KINETIC THEORY OF MATTER

Is it possible to explain the phenomena of heat in terms of the motions of particles interacting through simple forces? A closed vessel contains a certain mass of gas, air, for example, at a certain temperature. By heating we raise the temperature, and thus increase the energy. But how is this heat connected with motion? The possibility of such a connection is suggested both by our tentatively accepted philosophical point of view and by the way in which heat is generated by motion. Heat must be mechanical energy if every problem is a mechanical one. The object of the *kinetic theory* is to present the concept of matter just in this way. According to this theory a gas is a congregation of an enormous number of particles, or *molecules*, moving in all directions, colliding with each other and changing in direction of motion with each collision. There must exist an average speed of molecules, just as in a large human community there exists an average age, or an average wealth. There will therefore be an average kinetic energy per particle. More heat in the vessel means a greater average kinetic energy. Thus heat, according to this picture, is not a special form of energy different from the mechanical one but is just the kinetic energy of molecular motion. To any definite temperature there corresponds a definite average kinetic energy per molecule. This is, in fact, not an arbitrary assumption. We are forced to regard the kinetic energy of a molecule as a measure of the temperature of the gas if we wish to form a consistent mechanical picture of matter.

This theory is more than a play of the imagination. It can be shown that the kinetic theory of gases is not only in agreement with experiment, but actually leads to a more profound understanding of the facts. This may be illustrated by a few examples.

We have a vessel closed by a piston which can move freely. The vessel contains a certain amount of gas to be kept at a constant temperature. If the piston is initially at rest in some position it can be moved upward by removing and downward by adding weight. To push the piston down force must be used acting against the inner pressure of the gas. What is the mechanism of this inner pressure according to the kinetic theory? A tremendous number of particles constituting the gas are moving in all directions. They bombard the walls and the piston, bouncing back like balls thrown against a wall. This continual bombardment by a great number of particles keeps the piston at a certain height by opposing the force of gravity acting downward on the piston and the weights. In one direction there is a constant gravitational force, in the other very many irregular blows from the molecules. The net effect on the piston of all these small irregular forces must be equal to that of the force of gravity if there is to be equilibrium.

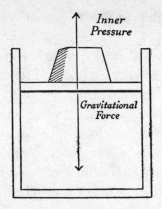

Suppose the piston were pushed down so as to compress the gas to a fraction of its former volume, say one-half, its temperature being kept unchanged. What, according to the kinetic theory, can we expect to happen? Will the force due to the bombardment be more or less effective than before? The particles are now packed more closely. Although the average kinetic energy is still the same, the collisions of the particles with the piston will now occur more frequently and thus the total force will be greater. It is clear from this picture presented by the kinetic theory that to keep the piston in this lower position more weight is required. This simple experimental fact is well known, but its prediction follows logically from the kinetic view of matter.

Consider another experimental arrangement. Take two vessels containing equal volumes of different gases, say hydrogen and nitrogen, both at the same temperature. Assume the two vessels are closed with identical pistons, on which are equal weights. This means, briefly, that both gases have the same volume, temperature, and pressure. Since the temperature is the same, so, according to the theory, is the average kinetic energy per particle. Since the pressures are equal, the two pistons are bombarded with the same total force. On the average every particle carries the same energy and both vessels have the same volume. Therefore, *the number of molecules in each must be the same*, although the gases are chemically different. This result is very important for the understanding of many chemical phenomena. It means that the number of molecules in a given volume, at a certain temperature and pressure, is something which is characteristic, not of a particular gas, but of all gases. It is most astonishing that the kinetic theory not only predicts the existence of such a universal number, but enables us to determine it. To this point we shall return very soon.

The kinetic theory of matter explains quantitatively as well as qualitatively the laws of gases as determined by experiment. Furthermore, the theory is not restricted to gases, although its greatest successes have been in this domain.

A gas can be liquefied by means of a decrease of temperature. A fall in the temperature of matter means a decrease in the average kinetic energy of its particles. It is therefore clear that the average kinetic energy of a liquid particle is smaller than that of a corresponding gas particle.

A striking manifestation of the motion of particles in liquids was given for the first time by the so-called *Brownian movement*, a remarkable phenomenon which would remain quite mysterious and incomprehensible without the kinetic theory of matter. It was first observed by the botanist Brown, and was explained eighty years later, at the beginning of this century. The only apparatus necessary for observing Brownian motion is a microscope, which need not even be a particularly good one.

Brown was working with grains of pollen of certain plants, that is:

particles or granules of unusually large size varying from one four-thousandth to about five-thousandth of an inch in length.

He reports further:

While examining the form of these particles immersed in water, I observed many of them evidently in motion. . . These motions were such as to satisfy me, after frequently repeated observation, that they arose neither from current in the fluid nor from its gradual evaporation, but belonged to the particle itself.

What Brown observed was the unceasing agitation of the granules when suspended in water and visible through the microscope. It is an impressive sight!

Is the choice of particular plants essential for the phenomenon? Brown answered this question by repeating the experiment with many different plants, and found that all the granules, if sufficiently small, showed such motion when suspended in water. Furthermore, he found the same kind of restless, irregular motion in very small particles of inorganic as well as organic substances. Even with a pulverized fragment of a sphinx he observed the same phenomenon!

How is this motion to be explained? It seems contradictory to all previous experience. Examination of the position of one suspended particle, say every thirty seconds, reveals the fantastic form of its path. The amazing thing is the apparently eternal character of the motion. A swinging pendulum placed in water soon comes to rest if not impelled by some external force. The existence of a never diminishing motion seems contrary to all experience. This difficulty was splendidly clarified by the kinetic theory of matter.

Looking at water through even our most powerful microscopes we cannot see molecules and their motion as pictured by the kinetic theory of matter. It must be concluded that if the theory of water as a congregation of particles is correct, the size of the particles must be beyond the limit of visibility of the best microscopes. Let us nevertheless stick to the theory and assume that it represents a consistent picture of reality. The Brownian particles visible through a microscope are bombarded by the smaller ones composing the water itself. The Brownian movement exists if the bombarded particles are sufficiently small. It exists because this bombardment is not uniform from all sides and cannot be averaged out, owing to its irregular and haphazard character. The observed motion is thus the result of the unobservable one. The behavior of the big particles reflects in some way that of the molecules, constituting, so to speak, a magnification so high that it becomes visible through the microscope. The irregular and haphazard character of the path of the Brownian particles reflects a similar irregularity in the path of the smaller particles which constitute matter. We can understand, therefore, that a quantitative study of Brownian movement can give us deeper insight into the kinetic theory of matter. It is apparent that the visible Brownian motion depends on the size of the invisible bombarding molecules. There would be no Brownian motion at all if the bombarding molecules did not possess a certain amount of energy or, in other words, if they did not have mass and velocity. That the study of Brownian motion can lead to a determination of the mass of a molecule is therefore not astonishing.

Through laborious research, both theoretical and experimental, the quantitative features of the kinetic theory were formed. The clew originating in the phenomenon of Brownian movement was one of those which led to the quantitative data. The same data can be obtained in different ways, starting from quite different clews. The fact that all these methods support the same view is most important, for it demonstrates the internal consistency of the kinetic theory of matter.

Only one of the many quantitative results reached by experiment and theory will be mentioned here. Suppose we have a gram of the lightest of all elements, hydrogen, and ask: how many particles are there in this one gram? The answer will characterize not only hydrogen but also all other gases, for we already know under what conditions two gases have the same number of particles.

The theory enables us to answer this question from certain measurements of the Brownian motion of a suspended particle. The answer is an astonishingly great number: a three followed by twenty-three other digits! The number of molecules in one gram of hydrogen is

303,000,000,000,000,000,000,000.

Imagine the molecules of a gram of hydrogen so increased in size that they are visible through a microscope, say that the diameter becomes one five-thousandth of an inch, such as that of a Brownian particle. Then, to pack them closely, we should have to use a box each side of which is about one-quarter of a mile long!

We can easily calculate the mass of one such hydrogen molecule by dividing 1 by the number quoted above. The answer is a fantastically small number:

0.000 000 000 000 000 000 000 0033 grams,

representing the mass of one molecule of hydrogen.

The experiments on Brownian motion are only some of the many independent experiments leading to the determination of this number which plays such an important part in physics.

In the kinetic theory of matter and in all its important achievements we see the realization of the general philosophical program: to reduce the explanation of all phenomena to the interaction between particles of matter.

WE SUMMARIZE:

In mechanics the future path of a moving body can be predicted and its past disclosed if its present condition and the forces acting upon it are known. Thus, for example, the future paths of all planets can be foreseen. The active

PLATE I

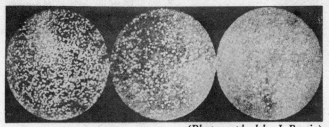

(Photographed by J. Perrin)

Brownian particles seen through a microscope.

(Photographed by Brumberg and Vavilov)

One Brownian particle photographed by a long exposure and covering a surface.

Consecutive positions observed for one of the Brownian particles.

The path averaged from these consecutive positions.

forces are Newton's gravitational forces depending on the distance alone. The great results of classical mechanics suggest that the mechanical view can be consistently applied to all branches of physics, that all phenomena can be explained by the action of forces representing either attraction or repulsion, depending only upon distance and acting between unchangeable particles.

In the kinetic theory of matter we see how this view, arising from mechanical problems, embraces the phenomena of heat and how it leads to a successful picture of the structure of matter.

II. THE DECLINE
OF THE MECHANICAL VIEW

The Decline of the Mechanical View

The two electric fluids ... The magnetic fluids ... The first serious difficulty ... The velocity of light ... Light as substance ... The riddle of color ... What is a wave? ... The wave theory of light ... Longitudinal or transverse light waves? ... Ether and the mechanical view

THE TWO ELECTRIC FLUIDS

THE following pages contain a dull report of some very simple experiments. The account will be boring not only because the description of experiments is uninteresting in comparison with their actual performance, but also because the meaning of the experiments does not become apparent until theory makes it so. Our purpose is to furnish a striking example of the role of theory in physics.

1. A metal bar is supported on a glass base, and each end of the bar is connected by means of a wire to an electroscope. What is an electroscope? It is a simple apparatus consisting essentially of two leaves of gold foil hanging from the end of a short piece of metal. This is enclosed in a glass jar or flask and the metal is in contact only with non-metallic bodies, called insulators. In addition to the electroscope and metal bar we are equipped with a hard rubber rod and a piece of flannel.

The experiment is performed as follows: we look

to see whether the leaves hang close together, for this is their normal position. If by chance they do not, a touch of the finger on the metal rod will bring them together. These preliminary steps being taken, the

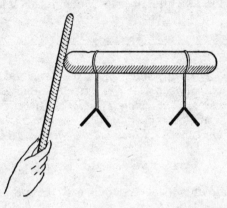

rubber rod is rubbed vigorously with the flannel and brought into contact with the metal. The leaves separate at once! They remain apart even after the rod is removed.

2. We perform another experiment, using the same apparatus as before, again starting with the gold leaves hanging close together. This time we do not bring the rubbed rod into actual contact with the metal, but only near it. Again the leaves separate. But there is a difference! When the rod is taken away without having touched the metal, the leaves immediately fall back to their normal position instead of remaining separated.

3. Let us change the apparatus slightly for a third experiment. Suppose that the metal bar consists of two pieces, joined together. We rub the rubber rod with flannel and again bring it near the metal. The same phenomenon occurs, the leaves separate. But now let

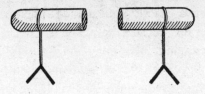

us divide the metal rod into its two separate parts and then take away the rubber rod. We notice that in this case the leaves remain apart, instead of falling back to their normal position as in the second experiment.

It is difficult to evince enthusiastic interest in these simple and naïve experiments. In the Middle Ages their performer would probably have been condemned; to us they seem both dull and illogical. It would be very difficult to repeat them, after reading the account only once, without becoming confused. Some notion of the theory makes them understandable. We could say more: it is hardly possible to imagine such experiments performed as accidental play, without the pre-existence of more or less definite ideas about their meaning.

We shall now point out the underlying ideas of a very simple and naïve theory which explains all the facts described.

There exist two *electric fluids*, one called *positive* (+) and the other *negative* (−). They are somewhat like substance in the sense already explained, in that the amount can be enlarged or diminished, but the total in any isolated system is preserved. There is, however, an essential difference between this case and that of heat, matter or energy. We have two electrical substances. It is impossible here to use the previous analogy of money unless it is somehow generalized. A body is electrically neutral if the positive and negative electric fluids exactly cancel each other. A man has nothing, either because he really has nothing, or because the amount of money put aside in his safe is exactly equal to the sum of his debts. We can compare the debit and credit entries in his ledger to the two kinds of electric fluids.

The next assumption of the theory is that two electric fluids of the same kind repel each other, while two of the opposite kind attract. This can be represented graphically in the following way:

$$\leftarrow \underset{+}{\bullet} \qquad \qquad \underset{+}{\bullet} \rightarrow$$

$$\leftarrow \underset{-}{\bullet} \qquad \qquad \underset{-}{\bullet} \rightarrow$$

$$\underset{+}{\bullet} \rightarrow \qquad \leftarrow \underset{-}{\bullet}$$

A final theoretical assumption is necessary: There are two kinds of bodies, those in which the fluids can move freely, called *conductors*, and those in which they cannot, called *insulators*. As is always true in such cases, this division is not to be taken too seriously. The ideal conductor or insulator is a fiction which can never be realized. Metals, the earth, the human body, are all examples of conductors, although not equally good. Glass, rubber, china, and the like are insulators. Air is only partially an insulator as everyone who has seen the described experiments knows. It is always a good excuse to ascribe the bad results of electrostatic experiments to the humidity of the air, which increases its conductivity.

These theoretical assumptions are sufficient to explain the three experiments described. We shall discuss them once more, in the same order as before, but in the light of the theory of electric fluids.

1. The rubber rod, like all other bodies under normal conditions, is electrically neutral. It contains the two fluids, positive and negative, in equal amounts. By rubbing with flannel we separate them. This statement is pure convention, for it is the application of the terminology created by the theory to the description of the process of rubbing. The kind of electricity that the rod has in excess afterwards is called negative, a name which is certainly only a matter of convention. If the experiments had been performed with a glass rod rubbed with cat's fur we should have had to call the excess positive, to conform with the accepted convention. To proceed with the experiment, we bring electric fluid to the metal conductor by touching it with the rubber. Here it moves freely, spreading over the whole metal including the gold leaves. Since the action of negative on negative is repulsion, the two leaves try to get as far from each other as possible and the result is the observed separation. The metal rests on glass or some other insulator so that the fluid remains on the conductor, as long as the conductivity of the air permits. We understand now why we have to touch the metal before beginning the experiment. In this case the metal, the human body, and the earth form one vast conductor, with the electric fluid so diluted that practically nothing remains on the electroscope.

2. This experiment begins just in the same way as the previous one. But instead of being allowed to touch the metal the rubber is now only brought near it. The

two fluids in the conductor, being free to move, are separated, one attracted and the other repelled. They mix again when the rubber rod is removed, as fluids of opposite kinds attract each other.

3. Now we separate the metal into two parts and afterwards remove the rod. In this case the two fluids cannot mix, so that the gold leaves retain an excess of one electric fluid and remain apart.

In the light of this simple theory all the facts mentioned here seem comprehensible. The same theory does more, enabling us to understand not only these, but many other facts in the realm of "electrostatics." The aim of every theory is to guide us to new facts, suggest new experiments, and lead to the discovery of new phenomena and new laws. An example will make this clear. Imagine a change in the second experiment. Suppose I keep the rubber rod near the metal and at the same time touch the conductor with my finger. What will happen now? Theory answers: the repelled fluid (—) can now make its escape through my body, with the result that only one fluid remains, the positive.

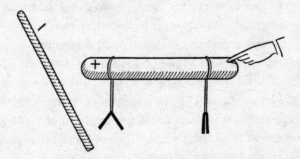

Only the leaves of the electroscope near the rubber rod will remain apart. An actual experiment confirms this prediction.

The theory with which we are dealing is certainly naïve and inadequate from the point of view of modern physics. Nevertheless it is a good example showing the characteristic features of every physical theory.

There are no eternal theories in science. It always happens that some of the facts predicted by a theory are disproved by experiment. Every theory has its period of gradual development and triumph, after which it may experience a rapid decline. The rise and fall of the substance theory of heat, already discussed here, is one of many possible examples. Others, more profound and important, will be discussed later. Nearly every great advance in science arises from a crisis in the old theory, through an endeavor to find a way out of the difficulties created. We must examine old ideas, old theories, although they belong to the past, for this is the only way to understand the importance of the new ones and the extent of their validity.

In the first pages of our book we compared the role of an investigator to that of a detective who, after gathering the requisite facts, finds the right solution by pure thinking. In one essential this comparison must be regarded as highly superficial. Both in life and in detective novels the crime is given. The detective must look for letters, fingerprints, bullets, guns, but at least he knows that a murder has been committed. This is not so for a scientist. It should not be difficult to imagine someone who knows absolutely nothing about electricity, since all the ancients lived happily enough without any knowledge of it. Let this man be given metal, gold foil, bottles, hard-rubber rod, flannel, in short, all the material required for performing our three experiments. He may be a very cultured person, but he will probably put wine into the bottles, use the flannel for cleaning, and never once entertain the idea of doing the things we have described. For the detective the crime is given, the problem formulated: who killed Cock Robin? The scientist must, at least in part, commit his own crime, as well as carry out the investigation. Moreover, his task is not to explain just one case, but all phenomena which have happened or may still happen.

In the introduction of the concept of fluids we see the influence of those mechanical ideas which attempt to explain everything by substances and simple forces acting between them. To see whether the mechanical point of view can be applied to the description of electrical phenomena, we must consider the following problem. Two small spheres are given, both with an electric charge, that is, both carrying an excess of one electric fluid. We know that the spheres will either attract or repel each other. But does the force depend only on the distance, and if so, how? The simplest guess seems to be that this force depends on the distance in the same way as gravitational force, which diminishes, say, to one-ninth of its former strength if the distance is made three times as great. The experiments performed by Coulomb showed that this law is really valid. A hundred years after Newton discovered the law of gravitation, Coulomb found a similar dependence of electrical force on distance. The two major differences between Newton's law and Coulomb's law are: gravitational attraction is always present, while electric forces exist only if the bodies possess electric charges. In the gravitational case there is only attraction, but electric forces may either attract or repel.

There arises here the same question which we considered in connection with heat. Are the electrical fluids weightless substances or not? In other words, is the weight of a piece of metal the same whether neutral or charged? Our scales show no difference. We conclude that the electric fluids are also members of the family of weightless substances.

Further progress in the theory of electricity requires the introduction of two new concepts. Again we shall avoid rigorous definitions, using instead analogies with concepts already familiar. We remember how essential it was for an understanding of the phenomena of heat to distinguish between heat itself and temperature. It is equally important here to distinguish between electric potential and electric charge. The difference between the two concepts is made clear by the analogy:

Electric potential — Temperature
Electric charge — Heat

Two conductors, for example two spheres of different size, may have the same electric charge, that is the same excess of one electric fluid, but the potential will be different in the two cases, being higher for the smaller and lower for the larger sphere. The electric fluid will have greater density and thus be more compressed on the small conductor. Since the repulsive forces must increase with the density, the tendency of the charge to escape will be greater in the case of the smaller sphere than in that of the larger. This tendency of charge to escape from a conductor is a direct measure of its potential. In order to show clearly the difference between charge and potential we shall formulate a few sentences describing the behavior of heated bodies, and the corresponding sentences concerning charged conductors.

HEAT	ELECTRICITY
Two bodies, initially at different temperatures, reach the same temperature after some time if brought into contact.	Two insulated conductors, initially at different electric potentials, very quickly reach the same potential if brought into contact.
Equal quantities of heat produce different changes of temperature in two bodies if their heat capacities are different.	Equal amounts of electric charge produce different changes of electric potential in two bodies if their electrical capacities are different.
A thermometer in contact with a body indicates—by the length of its mercury column—its own temperature and therefore the temperature of the body.	An electroscope in contact with a conductor indicates —by the separation of the gold leaves—its own electric potential and therefore the electric potential of the conductor.

But this analogy must not be pushed too far. An example shows the differences as well as the similarities. If a hot body is brought into contact with a cold one, the heat flows from the hotter to the colder. On the other hand suppose that we have two insulated conductors having equal but opposite charges, one positive and the other negative. The two are at different potentials. By convention we regard the potential corresponding to a negative charge as lower than that corresponding to a positive charge. If the two conductors are brought together or connected by a wire, it follows from the theory of electric fluids that they will show no charge and thus no difference of electric potential at all. We must imagine a "flow" of electric charge from one conductor to the other during the short time in which the potential difference is equalized. But how? Does the positive fluid flow to the negative body, or the negative fluid to the positive body?

In the material presented here we have no basis for deciding between these two alternatives. We can assume either of the two possibilities, or that the flow is simultaneous in both directions. It is only a matter of adopting a convention, and no significance can be attached to the choice, for we know no method of deciding the question experimentally. Further development leading to a much more profound theory of electricity gave an answer to this problem, which is quite meaningless when formulated in terms of the simple and primitive theory of electric fluids. Here we shall simply adopt the following mode of expression. The electric fluid flows from the conductor having the higher potential to that having the lower. In the case of our two conductors the electricity thus flows from positive to

negative. This expression is only a matter of convention and is at this point quite arbitrary. The whole difficulty indicates that the analogy between heat and electricity is by no means complete.

We have seen the possibility of adapting the mechanical view to a description of the elementary facts of electrostatics. The same is possible in the case of magnetic phenomena.

We shall proceed here in the same manner as before, starting with very simple facts and then seeking their theoretical explanation.

1. We have two long bar magnets, one suspended freely at its center, the other held in the hand. The ends of the two magnets are brought together in such a way that a strong attraction is noticed between them. This can always be done. If no attraction results we must turn the magnet and try the other end. Something will happen if the bars are magnetized at all. The

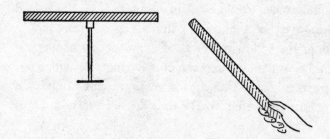

ends of the magnets are called their *poles*. To continue with the experiment we move the pole of the magnet held in the hand along the other magnet. A decrease in the attraction is noticed and when the pole reaches the middle of the suspended magnet there is no evidence of any force at all. If the pole is moved further in the same direction a repulsion is observed, attaining its greatest strength at the second pole of the hanging magnet.

2. The above experiment suggests another. Each magnet has two poles. Can we not isolate one of them? The idea is very simple: just break a magnet into two equal parts. We have seen that there is no force between the pole of one magnet and the middle of the other. But the result of actually breaking a magnet is surprising and unexpected. If we repeat the experiment described under (1), with only half a magnet suspended, the results are exactly the same as before! Where there was no trace of magnetic force previously, there is now a strong pole.

How are these facts to be explained? We can attempt to pattern a theory of magnetism after the theory of electric fluids. This is suggested by the fact that here, as in electrostatic phenomena, we have attraction and repulsion. Imagine two spherical conductors possessing equal charges, one positive and the other negative. Here "equal" means having the same absolute value; +5 and −5, for example, have the same absolute value. Let us assume that these spheres are connected by means of an insulator such as a glass rod.

Schematically this arrangement can be represented by an arrow directed from the negatively charged conductor to the positive one. We shall call the whole thing an electric *dipole*. It is clear that two such dipoles would behave exactly like the bar magnets in experiment (1). If we think of our invention as a model for a real magnet, we may say, assuming the existence of magnetic fluids, that a magnet is nothing but a *magnetic dipole*, having at its ends two fluids of different kinds. This simple theory, imitating the theory of electricity, is adequate for an explanation of the first experiment. There would be attraction at one end, repulsion at the other, and a balancing of equal and opposite forces in the middle. But what of the second experiment? By breaking the glass rod in the case of the electric dipole we get two isolated poles. The same ought to hold good for the iron bar of the magnetic dipole, contrary to the results of the second experiment. Thus this contradiction forces us to introduce a somewhat more subtle theory. Instead of our previous model we may imagine that the magnet consists of very small *elementary* magnetic dipoles which cannot be broken into separate poles. Order reigns in the magnet as a whole, for all the elementary dipoles are directed in the same way. We see immediately why cut-

ting a magnet causes two new poles to appear on the new ends, and why this more refined theory explains the facts of experiment (1) as well as (2).

For many facts, the simpler theory gives an explanation and the refinement seems unnecessary. Let us take an example: We know that a magnet attracts pieces of iron. Why? In a piece of ordinary iron the two magnetic fluids are mixed, so that no net effect results. Bringing a positive pole near acts as a "command of division" to the fluids, attracting the negative fluid of the iron and repelling the positive. The attraction between iron and magnet follows. If the magnet is removed, the fluids go back to more or less their original state, depending on the extent to which they remember the commanding voice of the external force.

Little need be said about the quantitative side of the problem. With two very long magnetized rods we could investigate the attraction (or repulsion) of their poles when brought near one another. The effect of the other ends of the rods is negligible if the rods are long enough. How does the attraction or repulsion depend on the distance between the poles? The answer given by Coulomb's experiment is that this dependence on distance is the same as in Newton's law of gravitation and Coulomb's law of electrostatics.

We see again in this theory the application of a general point of view: the tendency to describe all phenomena by means of attractive and repulsive forces depending only on distance and acting between unchangeable particles.

One well-known fact should be mentioned, for later we shall make use of it. The earth is a great magnetic dipole. There is not the slightest trace of an explanation as to why this is true. The North Pole is approximately the minus (−) and the South Pole the plus (+) magnetic pole of the earth. The names plus and minus are only a matter of convention, but when once fixed, enable us to designate poles in any other case. A magnetic needle supported on a vertical axis obeys the command of the magnetic force of the earth. It directs its (+) pole toward the North Pole, that is, toward the (−) magnetic pole of the earth.

Although we can consistently carry out the mechanical view in the domain of electric and magnetic phenomena introduced here, there is no reason to be particularly proud or pleased about it. Some features of the theory are certainly unsatisfactory if not discouraging. New kinds of substances had to be invented; two electric fluids and the elementary magnetic dipoles. The wealth of substances begins to be overwhelming!

The forces are simple. They are expressible in a similar way for gravitational, electric, and magnetic forces. But the price paid for this simplicity is high: the introduction of new weightless substances. These are rather artificial concepts, and quite unrelated to the fundamental substance, mass.

THE FIRST SERIOUS DIFFICULTY

We are now ready to note the first grave difficulty in the application of our general philosophical point of view. It will be shown later that this difficulty, together with another even more serious, caused a complete breakdown of the belief that all phenomena can be explained mechanically.

The tremendous development of electricity as a branch of science and technique began with the discovery of the electric current. Here we find in the history of science one of the very few instances in which accident seemed to play an essential role. The story of the convulsion of a frog's leg is told in many different ways. Regardless of the truth concerning details, there is no doubt that Galvani's accidental discovery led

Volta at the end of the eighteenth century to the construction of what is known as a *voltaic battery*. This is no longer of any practical use, but it still furnishes a very simple example of a source of current in school demonstrations and in textbook descriptions.

The principle of its construction is simple. There are several glass tumblers, each containing water with a little sulphuric acid. In each glass two metal plates, one copper and the other zinc, are immersed in the solution. The copper plate of one glass is connected to the zinc of the next, so that only the zinc plate of the first and the copper plate of the last glass remain unconnected. We can detect a difference in electric potential between the copper in the first glass and the zinc in the last by means of a fairly sensitive electroscope if the number of the "elements," that is, glasses with plates, constituting the battery, is sufficiently large.

It was only for the purpose of obtaining something easily measurable with apparatus already described that we introduced a battery consisting of several elements. For further discussion, a single element will serve just as well. The potential of the copper turns out to be higher than that of the zinc. "Higher" is used here in the sense in which +2 is greater than −2. If one conductor is connected to the free copper plate and another to the zinc, both will become charged, the first positively and the other negatively. Up to this point nothing particularly new or striking has appeared, and we may try to apply our previous ideas about potential differences. We have seen that a potential difference between two conductors can be quickly nullified by connecting them with a wire, so that there is a flow of electric fluid from one conductor to the other. This process was similar to the equalization of temperatures by heat flow. But does this work in the case of a voltaic battery? Volta wrote in his report that the plates behave like conductors:

. . . feebly charged, which act unceasingly or so that their charge after each discharge reestablishes itself; which, in a word, provides an unlimited charge or imposes a perpetual action or impulsion of the electric fluid.

The astonishing result of his experiment is that the potential difference between the copper and zinc plates does not vanish as in the case of two charged conductors connected by a wire. The difference persists, and according to the fluids theory it must cause a constant flow of electric fluid from the higher potential level (copper plate) to the lower (zinc plate). In an attempt to save the fluid theory, we may assume that some constant force acts to regenerate the potential difference and cause a flow of electric fluid. But the whole phenomenon is astonishing from the standpoint of energy. A noticeable quantity of heat is generated in the wire carrying the current, even enough to melt the wire if it is a thin one. Therefore, heat-energy is created in the wire. But the whole voltaic battery forms an isolated system, since no external energy is being supplied. If we want to save the law of conservation of energy we must find where the transformations take place, and at what expense the heat is created. It is not difficult to realize that complicated chemical processes are taking place in the battery, processes in which the immersed copper and zinc, as well as the liquid itself, take active parts. From the standpoint of energy this is the chain of transformations which are taking place: chemical energy → energy of the flowing electric fluid, i.e., the current → heat. A voltaic battery does not last forever; the chemical changes associated with the flow of electricity make the battery useless after a time.

The experiment which actually revealed the great difficulties in applying the mechanical ideas must sound strange to anyone hearing about it for the first time. It was performed by Oersted about a hundred and twenty years ago. He reports:

By these experiments it seems to be shown that the magnetic needle was moved from its position by help of a galvanic apparatus, and that, when the galvanic circuit was closed, but not when open, as certain very celebrated physicists in vain attempted several years ago.

Suppose we have a voltaic battery and a conducting wire. If the wire is connected to the copper plate but not to the zinc, there will exist a potential difference but no current can flow. Let us assume that the wire is bent to form a circle, in the center of which a magnetic needle is placed, both wire and needle lying in the same plane. Nothing happens as long as the wire does

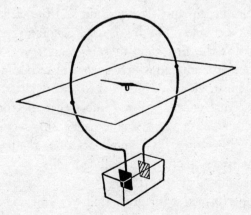

not touch the zinc plate. There are no forces acting, the existing potential difference having no influence whatever on the position of the needle. It seems difficult to understand why the "very celebrated physicists," as Oersted called them, expected such an influence.

But now let us join the wire to the zinc plate. Immediately a strange thing happens. The magnetic needle turns from its previous position. One of its poles now points to the reader if the page of this book represents the plane of the circle. The effect is that of a force, *perpendicular* to the plane, acting on the magnetic pole. Faced with the facts of the experiment, we can hardly avoid drawing such a conclusion about the direction of the force acting.

This experiment is interesting, in the first place, because it shows a relation between two apparently quite different phenomena, magnetism and electric current. There is another aspect even more important. The force between the magnetic pole and the small portions of the wire through which the current flows cannot lie along lines connecting the wire and needle, or the particles of flowing electric fluid and the elementary magnetic dipoles. The force is perpendicular to these lines! For the first time there appears a force quite different from that to which, according to our mechanical point of view, we intended to reduce all actions in the external world. We remember that the forces of gravitation, electrostatics, and magnetism, obeying the laws of Newton and Coulomb, act along the line joining the two attracting or repelling bodies.

This difficulty was stressed even more by an experiment performed with great skill by Rowland nearly sixty years ago. Leaving out technical details, this experiment could be described as follows. Imagine a small charged sphere. Imagine further that this sphere moves very fast in a circle at the center of which is a magnetic needle. This is, in principle, the same experiment as Oersted's, the only difference being that instead of an ordinary current we have a mechanically effected motion of the electric charge. Rowland found that the result is indeed similar to that observed when a current flows in a circular wire. The magnet is deflected by a perpendicular force.

Let us now move the charge faster. The force acting on the magnetic pole is, as a result, increased; the deflection from its initial position becomes more distinct. This observation presents another grave compli-

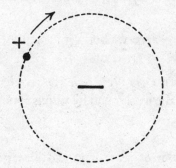

cation. Not only does the force fail to lie on the line connecting charge and magnet, but the intensity of the force depends on the velocity of the charge. The whole mechanical point of view was based on the belief that all phenomena can be explained in terms of forces depending only on the distance and not on the velocity. The result of Rowland's experiment certainly shakes this belief. Yet we may choose to be conservative and seek a solution within the frame of old ideas.

Difficulties of this kind, sudden and unexpected obstacles in the triumphant development of a theory, arise frequently in science. Sometimes a simple generalization of the old ideas seems, at least temporarily, to be a good way out. It would seem sufficient, in the present case, for example, to broaden the previous point of view and introduce more general forces between the elementary particles. Very often, however, it is impossible to patch up an old theory, and the difficulties result in its downfall and the rise of a new one. Here it was not only the behavior of a tiny magnetic needle which broke the apparently well-founded and successful mechanical theories. Another attack, even more vigorous, came from an entirely different angle. But this is another story, and we shall tell it later.

THE VELOCITY OF LIGHT

In Galileo's *Two New Sciences*, we listen to a conversation of the master and his pupils about the velocity of light:

SAGREDO: But of what kind and how great must we consider this speed of light to be? Is it instantaneous or momentary or does it, like other motion, require time? Can we not decide this by experiment?

SIMPLICIO: Everyday experience shows that the propagation of light is instantaneous; for when we see a piece of artillery fired, at great distance, the flash reaches our eyes without lapse of time; but the sound reaches the ear only after a noticeable interval.

SAGREDO: Well, Simplicio, the only thing I am able to infer from this familiar bit of experience is that sound, in reaching our ears, travels more slowly than light; it does not inform me whether the coming of the light is instantaneous or whether, although extremely rapid, it still occupies time. . . .

SALVIATI: The small conclusiveness of these and other similar observations once led me to devise a method by which one might accurately ascertain whether illumination, i.e., propagation of light, is really instantaneous. . . .

Salviati goes on to explain the method of his experiment. In order to understand his idea let us imagine that the velocity of light is not only finite, but also small, that the motion of light is slowed down, like that in a slow-motion film. Two men, A and B, have covered lanterns and stand, say, at a distance of one mile from each other. The first man, A, opens his lantern. The two have made an agreement that B will open his the moment he sees A's light. Let us assume that in our "slow motion" the light travels one mile in a second. A sends a signal by uncovering his lantern. B sees it after one second and sends an answering signal. This is received by A two seconds after he had sent his own. That is to say, if light travels with a speed of one mile per second, then two seconds will elapse between A's sending and receiving a signal, assuming that B is a mile away. Conversely, if A does not know the velocity of light but assumes that his companion kept the agreement, and he notices the opening of B's lantern two seconds after he opened his, he can conclude that the speed of light is one mile per second.

With the experimental technique available at that time Galileo had little chance of determining the velocity of light in this way. If the distance were a mile, he would have had to detect time intervals of the order of one hundred-thousandth of a second!

Galileo formulated the problem of determining the velocity of light, but did not solve it. The formulation of a problem is often more essential than its solution, which may be merely a matter of mathematical or experimental skill. To raise new questions, new possibilities, to regard old problems from a new angle, requires creative imagination and marks real advance in science. The principle of inertia, the law of conservation of energy were gained only by new and original thoughts about already well-known experiments and phenomena. Many instances of this kind will be found in the following pages of this book, where the importance of seeing known facts in a new light will be stressed and new theories described.

Returning to the comparatively simple question of determining the velocity of light, we may remark that it is surprising that Galileo did not realize that his experiment could be performed much more simply and accurately by one man. Instead of stationing his companion at a distance he could have mounted there a mirror, which would automatically send back the signal immediately after receiving it.

About two hundred and fifty years later this very principle was used by Fizeau, who was the first to determine the velocity of light by terrestrial experiments. It had been determined by Roemer much earlier, though less accurately, by astronomical observation.

It is quite clear that in view of its enormous magnitude, the velocity of light could be measured only by taking distances comparable to that between the earth and another planet of the solar system or by a great refinement of experimental technique. The first method was that of Roemer, the second that of Fizeau.

Since the days of these first experiments the very important number representing the velocity of light has been determined many times, with increasing accuracy. In our own century a highly refined technique was devised for this purpose by Michelson. The result of these experiments can be expressed simply: The velocity of light *in vacuo* is approximately 186,000 miles per second, or 300,000 kilometers per second.

LIGHT AS SUBSTANCE

Again we start with a few experimental facts. The number just quoted concerns the velocity of light *in vacuo*. Undisturbed, light travels with this speed through empty space. We can see through an empty glass vessel if we extract the air from it. We see planets, stars, nebulae, although the light travels from them to our eyes through empty space. The simple fact that we can see through a vessel whether or not there is air inside shows us that the presence of air matters very little. For this reason we can perform optical experiments in an ordinary room with the same effect as if there were no air.

One of the simplest optical facts is that the propagation of light is rectilinear. We shall describe a primitive and naïve experiment showing this. In front of a point source is placed a screen with a hole in it. A point source is a very small source of light, say, a small opening in a closed lantern. On a distant wall the hole in the screen will be represented as light on a dark background. The next drawing shows how this phenomenon is connected with the rectilinear propagation of light.

All such phenomena, even the more complicated cases in which light, shadow, and half-shadows appear, can be explained by the assumption that light, *in vacuo* or in air, travels along straight lines.

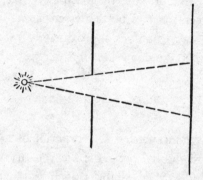

Let us take another example, a case in which light passes through matter. We have a light beam traveling through a vacuum and falling on a glass plate. What

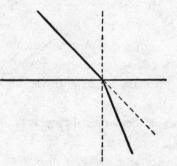

happens? If the law of rectilinear motion were still valid, the path would be that shown by the dotted line. But actually it is not. There is a break in the path, such as is shown in the drawing. What we observe here is the phenomenon known as *refraction*. The familiar appearance of a stick which seems to be bent in the mid-

dle if half-immersed in water is one of the many manifestations of refraction.

These facts are sufficient to indicate how a simple
mechanical theory of light could be devised. Our aim
here is to show how the ideas of substances, particles,
and forces penetrated the field of optics, and how
finally the old philosophical point of view broke down.

The theory here suggests itself in its simplest and
most primitive form. Let us assume that all lighted
bodies emit particles of light, or *corpuscles*, which, falling on our eyes, create the sensation of light. We are
already so accustomed to introduce new substances, if
necessary for a mechanical explanation, that we can do
it once more without any great hesitation. These corpuscles must travel along straight lines through empty
space with a known speed, bringing to our eyes messages from the bodies emitting light. All phenomena
exhibiting the rectilinear propagation of light support
the corpuscular theory, for just this kind of motion
was prescribed for the corpuscles. The theory also explains very simply the reflection of light by mirrors as
the same kind of reflection as that shown in the mechanical experiment of elastic balls thrown against a
wall, as the next drawing indicates.

The explanation of refraction is a little more difficult. Without going into details we can see the possibility of a mechanical explanation. If corpuscles fall on
the surface of glass, for example, it may be that a force
is exerted on them by the particles of the matter, a
force which strangely enough acts only in the immediate neighborhood of matter. Any force acting on a
moving particle changes the velocity, as we already
know. If the net force on the light-corpuscles is an

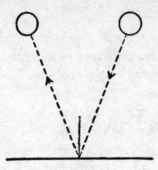

attraction perpendicular to the surface of the glass, the new motion will lie somewhere between the line of the original path and the perpendicular. This simple explanation seems to promise success for the corpuscular theory of light. To determine the usefulness and range of validity of the theory, however, we must investigate new and more complicated facts.

THE RIDDLE OF COLOR

It was again Newton's genius which explained for the first time the wealth of color in the world. Here is a description of one of Newton's experiments in his own words:

In the year 1666 (at which time I applied myself to the grinding of optick glasses of other figures than spherical) I procured me a triangular glass prism, to try therewith the celebrated phenomena of colours. And in order thereto, having darkened my chamber, and made a small hole in my window-shuts, to let in a convenient quantity of the sun's light, I placed my prism at its entrance, that it might thereby be refracted to the opposite wall. It was at first a very pleasing divertisement, to view the vivid and intense colours produced thereby.

The light from the sun is "white." After passing through a prism it shows all the colors which exist in the visible world. Nature herself reproduces the same result in the beautiful color scheme of the rainbow. Attempts to explain this phenomenon are very old. The Biblical story that a rainbow is God's signature to a covenant with man is, in a sense, a "theory." But it does not satisfactorily explain why the rainbow is repeated from time to time, and why always in connection with rain. The whole puzzle of color was first scientifically attacked and the solution pointed out in the great work of Newton.

One edge of the rainbow is always red and the other violet. Between them all other colors are arranged. Here is Newton's explanation of this phenomenon: every color is already present in white light. They all traverse interplanetary space and the atmosphere in unison and give the effect of white light. White light is, so to speak, a mixture of corpuscles of different kinds, belonging to different colors. In the case of Newton's experiment the prism separates them in space. According to the mechanical theory, refraction is due to forces acting on the particles of light and originating from the particles of glass. These forces are different for corpuscles belonging to different colors, being strongest for the violet and weakest for the red. Each of the colors will therefore be refracted along a different path and be separated from the others when the light leaves the prism. In the case of a rainbow, drops of water play the role of the prism.

The substance theory of light is now more complicated than before. We have not one light substance but many, each belonging to a different color. If, however, there is some truth in the theory, its consequences must agree with observation.

The series of colors in the white light of the sun, as revealed by Newton's experiment, is called the *spectrum* of the sun, or more precisely, its *visible spectrum*. The decomposition of white light into its components, as described here, is called the *dispersion* of light. The separated colors of the spectrum could be mixed together again by a second prism properly adjusted, unless the explanation given is wrong. The process should be just the reverse of the previous one. We should obtain white light from the previously separated colors. Newton showed by experiment that it is indeed possible to obtain white light from its spectrum and the spectrum from white light in this simple way as many times as one pleases. These experiments formed a strong support for the theory in which corpuscles belonging to each color behave as unchangeable substances. Newton wrote thus:

. . . which colours are not new generated, but only made apparent by being parted; for if they be again entirely mixt and blended together, they will again compose that colour, which they did before separation. And for the same reason, transmutations made by the convening of divers colours are not real; for when the difform rays are again severed, they will exhibit the very same colours which they did before they entered the composition; as you see blue and yellow powders, when finely mixed, appear to the naked eye, green, and yet the colours of the component corpuscles are not thereby really transmuted, but only blended. For when viewed with a good microscope they still appear blue and yellow interspersedly.

Suppose that we have isolated a very narrow strip of the spectrum. This means that of all the many colors we allow only one to pass through the slit, the others being stopped by a screen. The beam which comes through will consist of *homogeneous* light, that is, light which cannot be split into further components. This is a consequence of the theory and can be easily confirmed by experiment. In no way can such a beam of single color be divided further. There are simple means of obtaining sources of homogeneous light. For example, sodium, when incandescent, emits homogeneous yellow light. It is very often convenient to perform certain optical experiments with homogeneous light, since, as we can well understand, the result will be much simpler.

Let us imagine that suddenly a very strange thing happens: our sun begins to emit only homogeneous light of some definite color, say yellow. The great variety of colors on the earth would immediately vanish. Everything would be either yellow or black! This prediction is a consequence of the substance theory of light, for new colors cannot be created. Its validity can be confirmed by experiment: in a room where the only source of light is incandescent sodium everything is either yellow or black. The wealth of color in the world reflects the variety of color of which white light is composed.

The substance theory of light seems to work splendidly in all these cases, although the necessity for introducing as many substances as colors may make us somewhat uneasy. The assumption that all the corpuscles of light have exactly the same velocity in empty space also seems very artificial.

It is imaginable that another set of suppositions, a theory of entirely different character, would work just as well and give all the required explanations. Indeed, we shall soon witness the rise of another theory, based on entirely different concepts, yet explaining the same domain of optical phenomena. Before formulating the underlying assumptions of this new theory, however, we must answer a question in no way connected with these optical considerations. We must go back to mechanics and ask:

WHAT IS A WAVE?

A bit of gossip starting in Washington reaches New York very quickly, even though not a single individual who takes part in spreading it travels between these two cities. There are two quite different motions involved, that of the rumor, Washington to New York, and that of the persons who spread the rumor. The wind, passing over a field of grain, sets up a wave which spreads out across the whole field. Here again we must distinguish between the motion of the wave and the motion of the separate plants, which undergo only small oscillations. We have all seen the waves that spread in wider and wider circles when a stone is thrown into a pool of water. The motion of the wave is very different from that of the particles of water. The particles merely go up and down. The observed motion of the wave is that of a state of matter and not of matter itself. A cork floating on the wave shows this clearly, for it moves up and down in imitation of the actual motion of the water, instead of being carried along by the wave.

In order to understand better the mechanism of the wave let us again consider an idealized experiment.

Suppose that a large space is filled quite uniformly with water, or air, or some other "medium." Somewhere in the center there is a sphere. At the beginning of the experiment there is no motion at all. Suddenly the sphere begins to "breathe" rhythmically, expanding and contracting in volume, although retaining its spherical shape. What will happen in the medium? Let us begin our examination at the moment the sphere begins to expand. The particles of the medium in the immediate vicinity of the sphere are pushed out, so that the density of a spherical shell of water, or air, as the case may be, is increased above its normal value. Similarly, when the sphere contracts, the density of that part of the medium immediately surrounding it will be decreased. These changes of density are propagated throughout the entire medium. The particles constituting the medium perform only small vibrations, but the whole motion is that of a progressive wave. The essentially new thing here is that for the first time we consider the motion of something which is not matter, but energy propagated through matter.

Using the example of the pulsating sphere, we may introduce two general physical concepts, important for the characterization of waves. The first is the velocity with which the wave spreads. This will depend on the medium, being different for water and air, for example. The second concept is that of *wave-length*. In the case of waves on a sea or river it is the distance from the trough of one wave to that of the next, or from the crest of one wave to that of the next. Thus sea waves have greater wave-length than river waves. In the case of our waves set up by a pulsating sphere the wave-length is the distance, at some definite time, between two neighboring spherical shells showing maxima or minima of density. It is evident that this distance will not depend on the medium alone. The rate of pulsation of the sphere will certainly have a great effect, making the wave-length shorter if the pulsation becomes more rapid, longer if the pulsation becomes slower.

This concept of a wave proved very successful in physics. It is definitely a mechanical concept. The phenomenon is reduced to the motion of particles which, according to the kinetic theory, are constituents of matter. Thus every theory which uses the concept of wave can, in general, be regarded as a mechanical theory. For example, the explanation of acoustical phenomena is based essentially on this concept. Vibrating bodies, such as vocal cords and violin strings, are sources of sound waves which are propagated through the air in the manner explained for the pulsating sphere. It is thus possible to reduce all acoustical phenomena to mechanics by means of the wave concept.

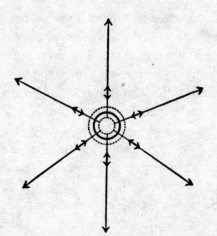

It has been emphasized that we must distinguish between the motion of the particles and that of the wave itself, which is a state of the medium. The two are very different but it is apparent that in our example of the pulsating sphere both motions take place in the same straight line. The particles of the medium oscillate along short line segments, and the density increases and decreases periodically in accordance with this motion. The direction in which the wave spreads and the line on which the oscillations lie are the same. This type of wave is called *longitudinal*. But is this the only kind of wave? It is important for our further considerations to realize the possibility of a different kind of wave, called *transverse*.

Let us change our previous example. We still have the sphere, but it is immersed in a medium of a different kind, a sort of jelly instead of air or water. Furthermore, the sphere no longer pulsates but rotates in one direction through a small angle and then back again, always in the same rhythmical way and about a definite

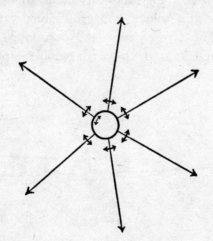

axis. The jelly adheres to the sphere and thus the adhering portions are forced to imitate the motion. These portions force those situated a little further away to imitate the same motion, and so on, so that a wave is set up in the medium. If we keep in mind the distinction between the motion of the medium and the motion of the wave we see that here they do not lie on the same line. The wave is propagated in the direction of the radius of the sphere, while the parts of the medium move perpendicularly to this direction. We have thus created a transverse wave.

Waves spreading on the surface of water are transverse. A floating cork only bobs up and down, but the wave spreads along a horizontal plane. Sound waves, on the other hand, furnish the most familiar example of longitudinal waves.

One more remark: the wave produced by a pulsating or oscillating sphere in a homogeneous medium is a *spherical* wave. It is called so because at any given moment all points on any sphere surrounding the source behave in the same way. Let us consider a portion of such a sphere at a great distance from the source. The farther away the portion is, and the smaller we take it, the more it resembles a plane. We

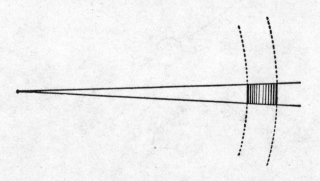

can say, without trying to be too rigorous, that there is no essential difference between a part of a plane and a part of a sphere whose radius is sufficiently large. We very often speak of small portions of a spherical wave far removed from the source as *plane waves*. The farther we place the shaded portion of our drawing from the center of the spheres and the smaller the angle between the two radii, the better our representation of a plane wave. The concept of a plane wave, like many other physical concepts, is no more than a fiction which can be realized with only a certain degree of accuracy. It is, however, a useful concept which we shall need later.

THE WAVE THEORY OF LIGHT

Let us recall why we broke off the description of optical phenomena. Our aim was to introduce another theory of light, different from the corpuscular one, but also attempting to explain the same domain of facts. To do this we had to interrupt our story and introduce the concept of waves. Now we can return to our subject.

It was Huygens, a contemporary of Newton, who put forward a quite new theory. In his treatise on light he wrote:

If, in addition, light takes time for its passage—which we are now going to examine—it will follow that this movement, impressed on the intervening matter, is successive; and consequently it spreads, as sound does, by spherical surfaces and waves, for I call them waves from their resemblance to those which are seen to be formed in water when a stone is thrown into it, and which present a successive spreading as circles, though these arise from another cause, and are only in a flat surface.

According to Huygens, light is a wave, a transference of energy and not of substance. We have seen that the corpuscular theory explains many of the observed facts. Is the wave theory also able to do this? We must again ask the questions which have already been answered by the corpuscular theory, to see whether the wave theory can do the answering just as well. We shall do this here in the form of a dialogue between N and H, where N is a believer in Newton's corpuscular theory, and H in Huygen's theory. Neither is allowed to use arguments developed after the work of the two great masters was finished.

N: In the corpuscular theory the velocity of light has a very definite meaning. It is the velocity at which the corpuscles travel through empty space. What does it mean in the wave theory?

H: It means the velocity of the light wave, of course. Every known wave spreads with some definite velocity, and so should a wave of light.

N: That is not as simple as it seems. Sound waves spread in air, ocean waves in water. Every wave must have a material medium in which it travels. But light passes through a vacuum, whereas sound does not. To assume a wave in empty space really means not to assume any wave at all.

H: Yes, that is a difficulty, although not a new one to me. My master thought about it very carefully, and decided that the only way out is to assume the existence of a hypothetical substance, the *ether*, a transparent medium permeating the entire universe. The universe is, so to speak, immersed in ether. Once we have the courage to introduce this concept, everything else becomes clear and convincing.

N: But I object to such an assumption. In the first place it introduces a new hypothetical substance, and we already have too many substances in physics. There is also another reason against it. You no doubt believe that we must explain everything in terms of mechanics. But what about the ether? Are you able to answer the simple question as to how the ether is constructed from its elementary particles and how it reveals itself in other phenomena?

H: Your first objection is certainly justified. But by introducing the somewhat artificial weightless ether we at once get rid of the much more artificial light corpuscles. We have only one "mysterious" substance instead of an infinite number of them corresponding to the great number of colors in the spectrum. Do you not think that this is real progress? At least all the difficulties are concentrated on one point. We no longer need the factitious assumption that particles belonging to different colors travel with the same speed through empty space. Your second argument is also true. We cannot give a mechanical explanation of ether. But there is no doubt that the future study of optical and perhaps other phenomena will reveal its structure. At present we must wait for new experiments and conclusions, but finally, I hope, we shall be able to clear up the problem of the mechanical structure of the ether.

N: Let us leave the question for the moment, since it cannot be settled now. I should like to see how your theory, even if we waive the difficulties, explains those phenomena which are so clear and understandable in the light of the corpuscular theory. Take, for example, the fact that light rays travel *in vacuo* or in air along straight lines. A piece of paper placed in front of a candle produces a distinct and sharply outlined shadow on the wall. Sharp shadows would not be possible if the wave theory of light were correct, for waves would bend around the edges of the paper and thus blur the shadow. A small ship is not an obstacle for waves on the sea, you know; they simply bend around it without casting a shadow.

H: That is not a convincing argument. Take short waves on a river impinging on the side of a large ship. Waves originating on one side of the ship will not be seen on the other. If the waves are small enough and the ship large enough a very distinct shadow appears. It is very probable that light seems to travel in straight lines only because its wave-length is very small in comparison with the size of ordinary obstacles and of apertures used in experiments. Possibly, if we could create a sufficiently small obstruction, no shadow would occur. We might meet with great experimental difficulties in constructing apparatus which would show whether light is capable of bending. Nevertheless, if such an experiment could be devised it would be crucial in deciding between the wave theory and the corpuscular theory of light.

N: The wave theory may lead to new facts in the future, but I do not know of any experimental data confirming it convincingly. Until it is definitely proved by experiment that light may be bent I do not see any reason for not believing in the corpuscular theory, which seems to me to be simpler, and therefore better, than the wave theory.

At this point we may interrupt the dialogue, though the subject is by no means exhausted.

It still remains to be shown how the wave theory explains the refraction of light and the variety of colors. The corpuscular theory is capable of this, as we know. We shall begin with refraction, but it will be useful to consider first an example having nothing to do with optics.

There is a large open space in which there are walking two men holding between them a rigid pole. At the beginning they are walking straight ahead, both with the same velocity. As long as their velocities remain the same, whether great or small, the stick will be undergoing parallel displacement; that is, it does not turn or change its direction. All consecutive positions of the pole are parallel to each other. But now imagine that for a time which may be as short as a fraction of a second the motions of the two men are not the same. What will happen? It is clear that during this moment the stick will turn, so that it will no longer be displaced parallel to its original position. When the equal velocities are resumed it is in a direction different from the previous one. This is shown clearly in the drawing.

The change in direction took place during the time

interval in which the velocities of the two walkers were different.

This example will enable us to understand the refraction of a wave. A plane wave traveling through the ether strikes a plate of glass. In the next drawing we see a wave which presents a comparatively wide front as it marches along. The wave front is a plane on which at any given moment all parts of the ether behave in precisely the same way. Since the velocity depends on the medium through which the light is passing it will be different in glass from the velocity in empty space. During the very short time in which the wave front enters the glass, different parts of the wave front will have different velocities. It is clear that the part which has reached the glass will travel with the velocity of light in glass, while the other still moves with the velocity of light in ether. Because of this difference in velocity along the wave front during the time of "immersion" in the glass, the direction of the wave itself will be changed.

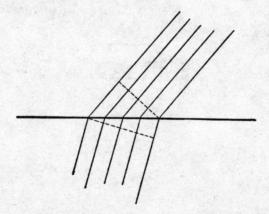

Thus we see that not only the corpuscular theory,

but also the wave theory, leads to an explanation of refraction. Further consideration, together with a little mathematics, shows that the wave theory explanation is simpler and better, and that the consequences are in perfect agreement with observation. Indeed, quantitative methods of reasoning enable us to deduce the velocity of light in a refractive medium if we know how the beam refracts when passing into it. Direct measurements splendidly confirm these predictions, and thus also the wave theory of light.

There still remains the question of color.

It must be remembered that a wave is characterized by two numbers, its velocity and its wave-length. The essential assumption of the wave theory of light is that *different wave-lengths correspond to different colors.* The wave-length of homogeneous yellow light differs from that of red or violet. Instead of the artificial segregation of corpuscles belonging to various colors we have the natural difference in wave-length.

It follows that Newton's experiments on the dispersion of light can be described in two different languages, that of the corpuscular theory and that of the wave theory. For example:

CORPUSCULAR LANGUAGE	WAVE LANGUAGE
The corpuscles belonging to different colors have the same velocity *in vacuo*, but different velocities in glass.	The rays of different wave-length belonging to different colors have the same velocity in the ether, but different velocities in glass.
White light is a composition of corpuscles belonging to different colors, whereas in the spectrum they are separated.	White light is a composition of waves of all wave-lengths, whereas in the spectrum they are separated.

It would seem wise to avoid the ambiguity resulting from the existence of two distinct theories of the same phenomena, by deciding in favor of one of them after a careful consideration of the faults and merits of each. The dialogue between N and H shows that this is no easy task. The decision at this point would be more a matter of taste than of scientific conviction. In Newton's time, and for more than a hundred years after, most physicists favored the corpuscular theory.

History brought in its verdict, in favor of the wave theory of light and against the corpuscular theory, at a much later date, the middle of the nineteenth century. In his conversation with H, N stated that a decision between the two theories was, in principle, experimentally possible. The corpuscular theory does not allow light to bend, and demands the existence of sharp shadows. According to the wave theory, on the other hand, a sufficiently small obstacle will cast no shadow. In the work of Young and Fresnel this result was experimentally realized and theoretical conclusions were drawn.

An extremely simple experiment has already been discussed, in which a screen with a hole was placed in front of a point source of light and a shadow appeared on the wall. We shall simplify the experiment further by assuming that the source emits homogeneous light. For the best results the source should be a strong one. Let us imagine that the hole in the screen is made smaller and smaller. If we use a strong source and succeed in making the hole small enough, a new and surprising phenomenon appears, something quite incomprehensible from the point of view of the corpus-

PLATE II

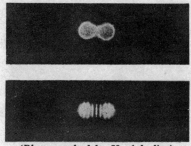

(*Photographed by V. Arkadiev*)

Above, we see a photograph of light spots after two beams have passed through two pin holes, one after the other. (One pin hole was opened, then covered and the other opened.) Below, we see stripes when light is allowed to pass through both pin holes simultaneously.

(*Photographed by V. Arkadiev*)

Diffraction of light bending around a small obstacle.

Diffraction of light passing through a small hole.

cular theory. There is no longer a sharp distinction between light and dark. Light gradually fades into the dark background in a series of light and dark rings. The appearance of rings is very characteristic of a wave theory. The explanation for alternating light and dark areas will be clear in the case of a somewhat different experimental arrangement. Suppose we have a sheet of dark paper with two pinholes through which light may pass. If the holes are close together and very small, and if the source of homogeneous light is strong enough, many light and dark bands will appear on the wall, gradually fading off at the sides into the dark background. The explanation is simple. A dark band is where a trough of a wave from one pinhole meets the crest of a wave from the other pinhole, so that the two cancel. A band of light is where two troughs or two crests from waves of the different pinholes meet and reinforce each other. The explanation is more complicated in the case of the dark and light rings of our previous example in which we used a screen with one hole, but the principle is the same. This appearance of dark and light stripes in the case of two holes and of light and dark rings in the case of one hole should be borne in mind, for we shall later return to a discussion of the two different pictures. The experiments described here show the *diffraction* of light, the deviation from the rectilinear propagation when small holes or obstacles are placed in the way of the light wave.

With the aid of a little mathematics we are able to go much further. It is possible to find out how great or, rather, how small the wave-length must be to produce a particular pattern. Thus the experiments described enable us to measure the wave-length of the homogeneous light used as a source. To give an idea of how small the numbers are we shall cite two wavelengths, those representing the extremes of the solar spectrum, that is, the red and the violet.

The wave-length of red light is 0.00008 cm.
The wave-length of violet light is 0.00004 cm.

We should not be astonished that the numbers are so small. The phenomenon of distinct shadow, that is, the phenomenon of rectilinear propagation of light, is observed in nature only because all apertures and obstacles ordinarily met with are extremely large in comparison with the wave-lengths of light. It is only when very small obstacles and apertures are used that light reveals its wave-like nature.

But the story of the search for a theory of light is by no means finished. The verdict of the nineteenth century was not final and ultimate. For the modern physicist the entire problem of deciding between corpuscles and waves again exists, this time in a much more profound and intricate form. Let us accept the defeat of the corpuscular theory of light until we recognize the problematic nature of the victory of the wave theory.

LONGITUDINAL OR TRANSVERSE LIGHT WAVES?

All the optical phenomena we have considered speak for the wave theory. The bending of light around small obstacles and the explanation of refraction are the strongest arguments in its favor. Guided by the mechanical point of view we realize that there is still one question to be answered: the determination of the mechanical properties of the ether. It is essential for the solution of this problem to know whether light waves in the ether are longitudinal or transverse. In other words: is light propagated like sound? Is the wave due to changes in the density of the medium, so that the oscillations of the particles are in the direction of the propagation? Or does the ether resemble an elastic jelly, a medium in which only transverse waves can be set up and whose particles move in a direction perpendicular to that in which the wave itself travels?

Before solving this problem let us try to decide which answer should be preferred. Obviously, we should be fortunate if light waves were longitudinal. The difficulties in designing a mechanical ether would be much simpler in this case. Our picture of ether might very probably be something like the mechanical picture of a gas that explains the propagation of sound waves. It would be much more difficult to form a picture of ether carrying transverse waves. To imagine a jelly as a medium made up of particles in such a way that transverse waves are propagated by means of it is no easy task. Huygens believed that the ether would turn out to be "air-like" rather than "jelly-like." But nature cares very little for our limitations. Was nature, in this case, merciful to the physicists attempting to understand all events from a mechanical point of view? In order to answer this question we must discuss some new experiments.

We shall consider in detail only one of many experiments which are able to supply us with an answer. Suppose we have a very thin plate of tourmaline crystal, cut in a particular way which we need not describe here. The crystal plate must be thin so that we are able to see a source of light through it. But now let us take two such plates and place both of them between our eyes and the light. What do we expect to see? Again a point of light, if the plates are sufficiently thin. The chances are very good that the experiment will confirm our expectation. Without worrying about the statement that it may be chance, let us assume we do see the light point through the two crystals. Now let us gradually change the position of one of the crystals by rotating it. This statement makes sense only if the position of the axis about which the rotation takes place is fixed. We shall take as an axis the line determined by the incoming ray. This means that we displace all the points of the one crystal except those

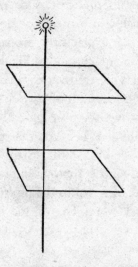

on the axis. A strange thing happens! The light gets weaker and weaker until it vanishes completely. It reappears as the rotation continues and we regain the initial view when the initial position is reached.

Without going into the details of this and similar experiments we can ask the following question: can these phenomena be explained if the light waves are longitudinal? In the case of longitudinal waves the particles of the ether would move along the axis, as the beam does. If the crystal rotates, nothing along the axis changes. The points on the axis do not move, and only a very small displacement takes place nearby. No such distinct change as the vanishing and appearance of a new picture could possibly occur for a longitudinal wave. This and many other similar phenomena can be explained only by the assumption that light waves are transverse and not longitudinal! Or, in other words, the "jelly-like" character of the ether must be assumed.

This is very sad! We must be prepared to face tremendous difficulties in the attempt to describe the ether mechanically.

ETHER AND THE MECHANICAL VIEW

The discussion of all the various attempts to understand the mechanical nature of the ether as a medium for transmitting light, would make a long story. A mechanical construction means, as we know, that the substance is built up of particles with forces acting along lines connecting them and depending only on the distance. In order to construct the ether as a jelly-like mechanical substance physicists had to make some highly artificial and unnatural assumptions. We shall not quote them here; they belong to the almost forgotten past. But the result was significant and important. The artificial character of all these assumptions, the necessity for introducing so many of them all quite independent of each other, was enough to shatter the belief in the mechanical point of view.

But there are other and simpler objections to ether than the difficulty of constructing it. Ether must be assumed to exist everywhere, if we wish to explain optical phenomena mechanically. There can be no empty space if light travels only in a medium.

Yet we know from mechanics that interstellar space does not resist the motion of material bodies. The planets, for example, travel through the ether-jelly without encountering any resistance such as a material medium would offer to their motion. If ether does not disturb matter in its motion, there can be no interaction between particles of ether and particles of matter. Light passes through ether and also through glass and water, but its velocity is changed in the latter substances. How can this fact be explained mechanically? Apparently only by assuming some interaction between ether particles and matter particles. We have just seen that in the case of freely moving bodies such interactions must be assumed not to exist. In other words, there is interaction between ether and matter in optical phenomena, but none in mechanical phenomena! This is certainly a very paradoxical conclusion!

There seems to be only one way out of all these difficulties. In the attempt to understand the phenomena of nature from the mechanical point of view, throughout the whole development of science up to the twentieth century, it was necessary to introduce artificial substances like electric and magnetic fluids, light corpuscles, or ether. The result was merely the concentration of all the difficulties in a few essential points, such as ether in the case of optical phenomena. Here all the fruitless attempts to construct an ether in some simple way, as well as the other objections, seem to indicate that the fault lies in the fundamental assumption that it is possible to explain all events in nature from a mechanical point of view. Science did not succeed in carrying out the mechanical program convincingly, and today no physicist believes in the possibility of its fulfillment.

In our short review of the principal physical ideas we have met some unsolved problems, have come upon difficulties and obstacles which discouraged the attempts to formulate a uniform and consistent view of all the phenomena of the external world. There was the unnoticed clew in classical mechanics of the equality of gravitational and inertial mass. There was the artificial character of the electric and magnetic fluids. There was, in the interaction between electric current and magnetic needle, an unsolved difficulty. It will be remembered that this force did not act in the line connecting the wire and the magnetic pole, and depended on the velocity of the moving charge. The law expressing its direction and magnitude was extremely complicated. And finally, there was the great difficulty with the ether.

Modern physics has attacked all these problems and solved them. But in the struggle for these solutions new and deeper problems have been created. Our knowledge is now wider and more profound than that of the physicist of the nineteenth century, but so are our doubts and difficulties.

WE SUMMARIZE:

In the old theories of electric fluids, in the corpuscular and wave theories of light, we witness the further attempts to apply the mechanical view. But in the realm of electric and optical phenomena we meet grave difficulties in this application.

A moving charge acts upon a magnetic needle. But the force, instead of depending only upon distance, depends also upon the velocity of the charge. The force neither repels nor attracts but acts perpendicular to the line connecting the needle and the charge.

In optics we have to decide in favor of the wave theory against the corpuscular theory of light. Waves spreading in a medium consisting of particles, with mechanical forces acting between them, are certainly a mechanical concept. But what is the medium through which light spreads and what are its mechanical properties? There is no hope of reducing the optical phenomena to the mechanical ones before this question is answered. But the difficulties in solving this problem are so great that we have to give it up and thus give up the mechanical view as well.

III. FIELD, RELATIVITY

Field, Relativity

THE FIELD AS REPRESENTATION

DURING the second half of the nineteenth century new
and revolutionary ideas were introduced into physics;
they opened the way to a new philosophical view, differing from the mechanical one. The results of the
work of Faraday, Maxwell, and Hertz led to the development of modern physics, to the creation of new concepts, forming a new picture of reality.

Our task now is to describe the break brought about
in science by these new concepts and to show how
they gradually gained clarity and strength. We shall
try to reconstruct the line of progress logically, without bothering too much about chronological order.

The new concepts originated in connection with the
phenomena of electricity, but it is simpler to introduce
them, for the first time, through mechanics. We know
that two particles attract each other and that this force
of attraction decreases with the square of the distance.

We can represent this fact in a new way, and shall do so even though it is difficult to understand the advan-

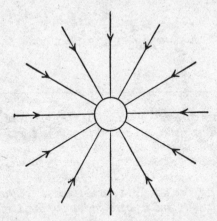

tage of this. The small circle in our drawing represents an attracting body, say, the sun. Actually, our diagram should be imagined as a model in space and not as a drawing on a plane. Our small circle, then, stands for a sphere in space, say, the sun. A body, the so-called *test body*, brought somewhere within the vicinity of the sun will be attracted along the line connecting the centers of the two bodies. Thus the lines in our drawing indicate the direction of the attracting force of the sun for different positions of the test body. The arrow on each line shows that the force is directed toward the sun; this means the force is an attraction. These are the *lines of force of the gravitational field.* For the moment, this is merely a name and there is no reason for stressing it further. There is one characteristic feature of our drawing which will be emphasized later. The lines of force are constructed in space, where no matter is present. For the moment, all the lines of force, or briefly speaking, the *field*, indicate only how a test body would behave if brought into the vicinity of the sphere for which the field is constructed.

The lines in our space model are always perpendicular to the surface of the sphere. Since they diverge from one point, they are dense near the sphere and become less and less so farther away. If we increase the distance from the sphere twice or three times, then the density of the lines, in our space-model, though not in the drawing, will be four or nine times less. Thus the lines serve a double purpose. On the one hand they show the direction of the force acting on a body brought into the neighborhood of the sphere-sun. On the other hand the density of the lines of force in space shows how the force varies with the distance. The drawing of the field, correctly interpreted, represents the direction of the gravitational force and its dependence on distance. One can read the law of gravitation from such a drawing just as well as from a description of the action in words, or in the precise and economical language of mathematics. This *field representation*, as we shall call it, may appear clear and interesting but there is no reason to believe that it marks any real advance. It would be quite difficult to prove its usefulness in the case of gravitation. Some may, perhaps, find it helpful to regard these lines as something more than drawings, and to imagine the real actions of force passing through them. This may be done, but then the speed of the actions along the lines of force must be assumed as infinitely great! The force between two bodies, according to Newton's law, depends only on distance; time does not enter the picture. The force has to pass from one body to another in no time! But, as motion with infinite speed cannot mean much to any reasonable person, an attempt to make our drawing something more than a model leads nowhere.

We do not intend, however, to discuss the gravitational problem just now. It served only as an introduction, simplifying the explanation of similar methods of reasoning in the theory of electricity.

We shall begin with a discussion of the experiment which created serious difficulties in our mechanical interpretation. We had a current flowing through a wire circuit in the form of a circle. In the middle of the circuit was a magnetic needle. The moment the current began to flow a new force appeared, acting on the magnetic pole, and perpendicular to any line connecting the wire and the pole. This force, if caused by a circulating charge, depended, as shown by Rowland's experiment, on the velocity of the charge. These experimental facts contradicted the philosophical view that all forces must act on the line connecting the particles and can depend only upon distance.

The exact expression for the force of a current acting on a magnetic pole is quite complicated, much more so, indeed, than the expression for gravitational forces. We can, however, attempt to visualize the actions just as we did in the case of a gravitational force. Our question is: with what force does the current act upon a magnetic pole placed somewhere in its vicinity? It would be rather difficult to describe this force in words. Even a mathematical formula would be complicated and awkward. It is best to represent all we know about the acting forces by a drawing, or rather by a spatial model, with lines of force. Some difficulty is caused by the fact that a magnetic pole exists only in connection with another magnetic pole, form-

ing a dipole. We can, however, always imagine the
magnetic needle of such length that only the force acting upon the pole nearer the current has to be taken
into account. The other pole is far enough away for
the force acting upon it to be negligible. To avoid ambiguity we shall say that the magnetic pole brought
nearer to the wire is the *positive* one.

The character of the force acting upon the positive
magnetic pole can be read from our drawing.

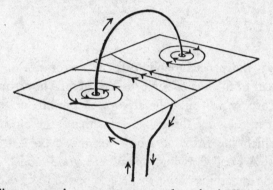

First we notice an arrow near the wire indicating the
direction of the current, from higher to lower poten
tial. All other lines are just lines of force belonging to
this current and lying on a certain plane. If drawn
properly, they tell us the direction of the force vector
representing the action of the current on a given positive magnetic pole as well as something about the
length of this vector. Force, as we know, is a vector
and to determine it we must know its direction as well
as its length. We are chiefly concerned with the problem of the direction of the force acting upon a pole.
Our question is: how can we find, from the drawing,
the direction of the force, at any point in space?

The rule for reading the direction of a force from

such a model is not as simple as in our previous example, where the lines of force were straight. In our next
diagram only one line of force is drawn in order to

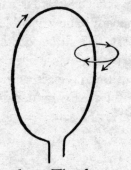

clarify the procedure. The force vector lies on the
tangent to the line of force, as indicated. The arrow of
the force vector and the arrows on the line of force
point in the same direction. Thus this is the direction
in which the force acts on a magnetic pole at this
point. A good drawing, or rather a good model, also
tells us something about the length of the force vector
at any point. This vector has to be longer where the
lines are denser, i.e., near the wire, shorter where the
lines are less dense, i.e., far from the wire.

In this way, the lines of force, or in other words, the
field, enable us to determine the forces acting on a
magnetic pole at any point in space. This, for the time
being, is the only justification for our elaborate construction of the field. Knowing what the field expresses, we shall examine with a far deeper interest the
lines of force corresponding to the current. These lines
are circles surrounding the wire and lying on the plane
perpendicular to that in which the wire is situated.
Reading the character of the force from the drawing

we come once more to the conclusion that the force acts in a direction perpendicular to any line connecting the wire and the pole, for the tangent to a circle is always perpendicular to its radius. Our entire knowledge of the acting forces can be summarized in the construction of the field. We sandwich the concept of the field between that of the current and that of the magnetic pole in order to represent the acting forces in a simple way.

Every current is associated with a magnetic field, i.e., a force always acts on a magnetic pole brought near the wire through which a current flows. We may remark in passing that this property enables us to construct sensitive apparatus for detecting the existence of a current. Once having learned how to read the character of the magnetic forces from the field model of a current, we shall always draw the field surrounding the wire through which the current flows, in order to represent the action of the magnetic forces at any point in space. Our first example is the so-called solenoid. This is, in fact, a coil of wire as shown in the drawing. Our aim is to learn, by experiment, all we can about the magnetic field associated with the current flowing through a solenoid and to incorporate this knowledge in the construction of a field. A drawing represents our result. The curved lines of force are closed, and surround the solenoid in a way characteristic of the magnetic field of a current. See top p. 132.

The field of a bar magnet can be represented in the same way as that of a current. Another drawing shows this. The lines of force are directed from the positive to the negative pole. The force vector always lies on

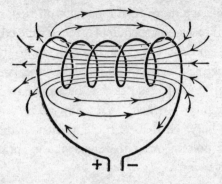

the tangent to the line of force and is longest near the poles because the density of the lines is greatest at these points. The force vector represents the action of the magnet on a positive magnetic pole. In this case the magnet and not the current is the "source" of the field.

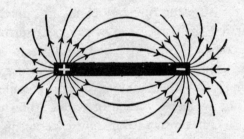

Our last two drawings should be carefully compared. In the first, we have the magnetic field of a current flowing through a solenoid; in the second, the field of a bar magnet. Let us ignore both the solenoid and the bar and observe only the two outside fields. We immediately notice that they are of exactly the same character; in each case the lines of force lead from one end of the solenoid or bar to the other.

The field representation yields its first fruit! It would be rather difficult to see any strong similarity between the current flowing through a solenoid and a bar magnet if this were not revealed by our construc- tion of the field.

The concept of field can now be put to a much more severe test. We shall soon see whether it is any- thing more than a new representation of the acting forces. We could reason: assume, for a moment, that the field characterizes all actions determined by its sources in a unique way. This is only a guess. It would mean that if a solenoid and a bar magnet have the same field, then all their influences must also be the same. It would mean that two solenoids, carrying electric cur- rents, behave like two bar magnets, that they attract or repel each other depending, exactly as in the case of bars, on their relative positions. It would also mean that a solenoid and a bar attract or repel each other in the same way as two bars. Briefly speaking, it would mean that all actions of a solenoid through which a current flows, and of a corresponding bar magnet are the same, since the field alone is responsible for them, and the field in both cases is of the same character. Experiment fully confirms our guess!

How difficult it would be to find those facts without the concept of field! The expression for a force acting between a wire through which a current flows and a magnetic pole is very complicated. In the case of two solenoids we should have to investigate the forces with which two currents act upon each other. But if we do this, with the help of the field, we immediately notice the character of all those actions at the moment when the similarity between the field of a solenoid and that of a bar magnet is seen.

We have the right to regard the field as something

much more than we did at first. The properties of the field alone appear to be essential for the description of phenomena; the differences in source do not matter. The concept of field reveals its importance by leading to new experimental facts.

The field proved a very helpful concept. It began as something placed between the source and the magnetic needle in order to describe the acting force. It was thought of as an "agent" of the current, through which all action of the current was performed. But now the agent also acts as an interpreter, one who translates the laws into a simple, clear language, easily understood.

The first success of the field description suggests that it may be convenient to consider all actions of currents, magnets and charges indirectly, i.e., with the help of the field as an interpreter. A field may be regarded as something always associated with a current. It is there even in the absence of a magnetic pole to test its existence. Let us try to follow this new clew consistently.

The field of a charged conductor can be introduced in much the same way as the gravitational field, or the field of a current or magnet. Again only the simplest example! To design the field of a positively charged sphere, we must ask what kind of forces are acting on a small positively charged test body brought near the source of the field, the charged sphere. The fact that we use a positively and not a negatively charged test body is merely a convention, indicating in which direction the arrows on the line of force should be drawn. The model is analogous to that of a gravitational field (p. 126) because of the similarity between Coulomb's law and Newton's. The only difference between the

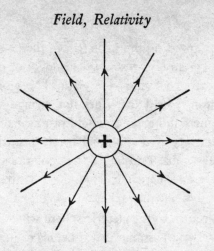

two models is that the arrows point in opposite directions. Indeed, we have repulsion of two positive
charges and attraction of two masses. However, the
field of a sphere with a negative charge will be identical with a gravitational field since the small positive
testing charge will be attracted by the source of the
field.

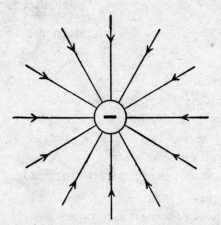

If both electric and magnetic poles are at rest, there
is no action between them, neither attraction nor re-

pulsion. Expressing the same fact in the field language we can say: an electrostatic field does not influence a magnetostatic one and vice versa. The words "static field" mean a field that does not change with time. The magnets and charges would rest near one another for an eternity if no external forces disturbed them. Electrostatic, magnetostatic and gravitational fields are all of different character. They do not mix; each preserves its individuality regardless of the others.

Let us return to the electric sphere which was, until now, at rest, and assume that it begins to move due to the action of some external force. The charged sphere moves. In the field language this sentence reads: the field of the electric charge changes with time. But the motion of this charged sphere is, as we already know from Rowland's experiment, equivalent to a current. Further, every current is accompanied by a magnetic field. Thus the chain of our argument is:

motion of charge → change of an electric field
↓
current → associated magnetic field.

We, therefore, conclude: *The change of an electric field produced by the motion of a charge is always accompanied by a magnetic field.*

Our conclusion is based on Oersted's experiment but it covers much more. It contains the recognition that the association of an electric field, changing in time, with a magnetic field is essential for our further argument.

As long as a charge is at rest there is only an electrostatic field. But a magnetic field appears as soon as the charge begins to move. We can say more. The magnetic field created by the motion of the charge will be stronger if the charge is greater and if it moves faster. This also is a consequence of Rowland's experiment. Once again using the field language, we can say: the faster the electric field changes, the stronger the accompanying magnetic field.

We have tried here to translate familiar facts from the language of fluids, constructed according to the old mechanical view, into the new language of fields. We shall see later how clear, instructive, and far-reaching our new language is.

THE TWO PILLARS OF THE FIELD THEORY

"The change of an electric field is accompanied by a magnetic field." If we interchange the words "magnetic" and "electric," our sentence reads: "The change of a magnetic field is accompanied by an electric field." Only an experiment can decide whether or not this statement is true. But the idea of formulating this problem is suggested by the use of the field language.

Just over a hundred years ago, Faraday performed an experiment which led to the great discovery of induced currents.

The demonstration is very simple. We need only a solenoid or some other circuit, a bar magnet, and one of the many types of apparatus for detecting the existence of an electric current. To begin with, a bar magnet is kept at rest near a solenoid which forms a closed circuit. No current flows through the wire, for no source is present. There is only the magnetostatic field of the bar magnet which does not change with time. Now, we quickly change the position of the magnet either by removing it or by bringing it nearer the solenoid, whichever we prefer. At this moment, a current will appear for a very short time and then vanish.

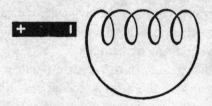

Whenever the position of the magnet is changed, the current reappears, and can be detected by a sufficiently sensitive apparatus. But a current—from the point of view of the field theory—means the existence of an electric field forcing the flow of the electric fluids through the wire. The current, and therefore the electric field, too, vanishes when the magnet is again at rest.

Imagine for a moment that the field language is unknown and the results of this experiment have to be described, qualitatively and quantitatively, in the language of old mechanical concepts. Our experiment then shows: by the motion of a magnetic dipole a new force was created, moving the electric fluid in the wire. The next question would be: upon what does this force depend? This would be very difficult to answer. We should have to investigate the dependence of the force upon the velocity of the magnet, upon its shape, and upon the shape of the circuit. Furthermore, this experiment, if interpreted in the old language, gives us no hint at all as to whether an induced current can be excited by the motion of another circuit carrying a current, instead of by motion of a bar magnet.

It is quite a different matter if we use the field language and again trust our principle that the action is determined by the field. We see at once that a solenoid through which a current flows would serve as well as a bar magnet. The drawing shows two solenoids: one, small, through which a current flows, and the other, in which the induced current is detected, larger. We

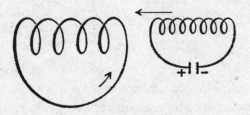

could move the small solenoid, as we previously moved the bar magnet, creating an induced current in the larger solenoid. Furthermore, instead of moving the small solenoid, we could create and destroy a magnetic field by creating and destroying the current, that is, by opening and closing the circuit. Once again, new facts suggested by the field theory are confirmed by experiment!

Let us take a simpler example. We have a closed wire without any source of current. Somewhere in the vicinity is a magnetic field. It means nothing to us whether the source of this magnetic field is another circuit through which an electric current flows, or a bar magnet. Page 140 shows the closed circuit and the magnetic lines of force. The qualitative and quantitative description of the induction phenomena is very simple in terms of the field language. As marked on the drawing, some lines of force go through the surface bounded by the wire. We have to consider the lines of force cutting that part of the plane which has the wire

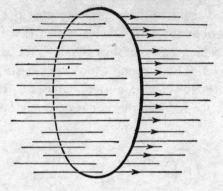

for a rim. No electric current is present so long as the field does not change, no matter how great its strength. But a current begins to flow through the rim-wire as soon as the number of lines passing through the surface surrounded by wire changes. The current is determined by the change, however it may be caused, of the number of lines passing the surface. This change in the number of lines of force is the only essential concept for both the qualitative and the quantitative descriptions of the induced current. "The number of lines changes" means that the density of the lines changes and this, we remember, means that the field strength changes.

These then are the essential points in our chain of reasoning: change of magnetic field → induced current → motion of charge → existence of an electric field.

Therefore: *a changing magnetic field is accompanied by an electric field*.

Thus we have found the two most important pillars of support for the theory of the electric and magnetic field. The first is the connection between the changing electric field and the magnetic field. It arose from Oersted's experiment on the deflection of a magnetic needle and led to the conclusion: *a changing electric field is accompanied by a magnetic field.*

The second connects the changing magnetic field with the induced current and arose from Faraday's experiment. Both formed a basis for quantitative description.

Again the electric field accompanying the changing magnetic field appears as something real. We had to imagine, previously, the magnetic field of a current existing without the testing pole. Similarly, we must claim here that the electric field exists without the wire testing the presence of an induced current.

In fact, our two-pillar structure could be reduced to only one, namely, to that based on Oersted's experiment. The result of Faraday's experiment could be deduced from this with the law of conservation of energy. We used the two-pillar structure only for the sake of clearness and economy.

One more consequence of the field description should be mentioned. There is a circuit through which a current flows, with for instance, a voltaic battery as the source of the current. The connection between the wire and the source of the current is suddenly broken. There is, of course, no current now! But during this short interruption an intricate process takes place, a process which could again have been foreseen by the field theory. Before the interruption of the current there was a magnetic field surrounding the wire. This ceased to exist the moment the current was interrupted. Therefore, through the interruption of a current, a magnetic field disappeared. The number of lines of force passing through the surface surrounded by the wire changed very rapidly. But such a rapid change, however it is produced, must create an induced current. What really matters is the change of the magnetic field making the induced current stronger if the change is greater. This consequence is another test for the theory. The disconnection of a current must be accompanied by the appearance of a strong, momentary induced current. Experiment again confirms the prediction. Anyone who has ever disconnected a current must have noticed that a spark appears. This spark reveals the strong potential differences caused by the rapid change of the magnetic field.

The same process can be looked at from a different point of view, that of energy. A magnetic field disappeared and a spark was created. A spark represents energy, therefore, so also must the magnetic field. To use the field concept and its language consistently, we must regard the magnetic field as a store of energy. Only in this way shall we be able to describe the electric and magnetic phenomena in accordance with the law of conservation of energy.

Starting as a helpful model the field became more and more real. It helped us to understand old facts and led us to new ones. The attribution of energy to the field is one step further in the development in which the field concept was stressed more and more, and the concepts of substances, so essential to the mechanical point of view, were more and more suppressed.

THE REALITY OF THE FIELD

The quantitative, mathematical description of the laws of the field is summed up in what are called Maxwell's equations. The facts mentioned so far led to the formulation of these equations but their content is much richer than we have been able to indicate. Their simple form conceals a depth revealed only by careful study.

The formulation of these equations is the most important event in physics since Newton's time, not only because of their wealth of content, but also because they form a pattern for a new type of law.

The characteristic features of Maxwell's equations, appearing in all other equations of modern physics, are summarized in one sentence. Maxwell's equations are laws representing the *structure* of the field.

Why do Maxwell's equations differ in form and character from the equations of classical mechanics? What does it mean that these equations describe the structure of the field? How is it possible that, from the results of Oersted's and Faraday's experiments, we can form a new type of law, which proves so important for the further development of physics?

We have already seen, from Oersted's experiment, how a magnetic field coils itself around a changing electric field. We have seen, from Faraday's experiment, how an electric field coils itself around a changing magnetic field. To outline some of the characteristic features of Maxwell's theory, let us, for the moment, focus all our attention on one of these experiments, say, on that of Faraday. We repeat the drawing in which an electric current is induced by a changing magnetic field. We already know that an induced current appears if the number of lines of force, passing the surface bounded by the wire, changes. Then the current

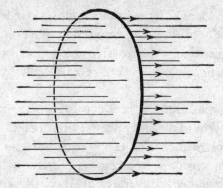

will appear if the magnetic field changes or the circuit is deformed or moved: if the number of magnetic lines passing through the surface is changed, no matter how this change is caused. To take into account all these various possibilities, to discuss their particular influences, would necessarily lead to a very complicated theory. But can we not simplify our problem? Let us try to eliminate from our considerations everything which refers to the shape of the circuit, to its length, to the surface enclosed by the wire. Let us imagine that the circuit in our last drawing becomes smaller and smaller, shrinking gradually to a very small circuit enclosing a certain point in space. Then everything concerning shape and size is quite irrelevant. In this limiting process where the closed curve shrinks to a point, size and shape automatically vanish from our considerations and we obtain laws connecting changes of magnetic and electric field at an arbitrary point in space at an arbitrary instant.

Thus, this is one of the principal steps leading to Maxwell's equations. It is again an idealized experiment performed in imagination by repeating Faraday's experiment with a circuit shrinking to a point.

We should really call it half a step rather than a whole one. So far our attention has been focused on Faraday's experiment. But the other pillar of the field theory, based on Oersted's experiment, must be considered just as carefully and in a similar manner. In this experiment the magnetic lines of force coil themselves around the current. By shrinking the circular magnetic lines of force to a point, the second half-step is performed and the whole step yields a connection between the changes of the magnetic and electric fields at an arbitrary point in space and at an arbitrary instant.

But still another essential step is necessary. According to Faraday's experiment, there must be a wire testing the existence of the electric field, just as there must be a magnetic pole, or needle, testing the existence of a magnetic field in Oersted's experiment. But Maxwell's new theoretical idea goes beyond these experimental facts. The electric and magnetic field, or in short, the *electromagnetic* field is, in Maxwell's theory, something real. The electric field is produced by a changing magnetic field, quite independently, whether or not there is a wire to test its existence; a magnetic field is produced by a changing electric field, whether or not there is a magnetic pole to test its existence.

Thus two essential steps led to Maxwell's equations. The first: in considering Oersted's and Rowland's experiments, the circular line of the magnetic field coiling itself around the current and the changing electric field, had to be shrunk to a point; in considering Faraday's experiment, the circular line of the electric field coiling itself around the changing magnetic field had to be shrunk to a point. The second step consists of the realization of the field as something real; the electromagnetic field once created exists, acts, and changes according to Maxwell's laws.

Maxwell's equations describe the structure of the electromagnetic field. All space is the scene of these laws and not, as for mechanical laws, only points in which matter or charges are present.

We remember how it was in mechanics. By knowing the position and velocity of a particle at one single instant, by knowing the acting forces, the whole future path of the particle could be forseen. In Maxwell's theory, if we know the field at one instant only, we can deduce from the equations of the theory how the whole field will change in space and time. Maxwell's equations enable us to follow the history of the field, just as the mechanical equations enabled us to follow the history of material particles.

But there is still one essential difference between mechanical laws and Maxwell's laws. A comparison of Newton's gravitational laws and Maxwell's field laws will emphasize some of the characteristic features expressed by these equations.

With the help of Newton's laws we can deduce the motion of the earth from the force acting between the sun and the earth. The laws connect the motion of the earth with the action of the far-off sun. The earth and the sun, though so far apart, are both actors in the play of forces.

In Maxwell's theory there are no material actors. The mathematical equations of this theory express the laws governing the electromagnetic field. They do not, as in Newton's laws, connect two widely separated events; they do not connect the happenings *here* with the conditions *there*. The field *here* and *now* depends on the field in the *immediate neighborhood* at a time *just past*. The equations allow us to predict what will happen a little further in space and a little later in time, if we know what happens here and now. They allow us to increase our knowledge of the field by small steps. We can deduce what happens here from that which happened far away by the summation of these very small steps. In Newton's theory, on the contrary, only big steps connecting distant events are permissible. The experiments of Oersted and Faraday can be regained from Maxwell's theory, but only by the summation of small steps each of which is governed by Maxwell's equations.

A more thorough mathematical study of Maxwell's equations shows that new and really unexpected conclusions can be drawn and the whole theory submitted to a test on a much higher level, because the theoretical consequences are now of a quantitative character and are revealed by a whole chain of logical arguments.

Let us again imagine an idealized experiment. A small sphere with an electric charge is forced, by some external influence, to oscillate rapidly and in a rhythmical way, like a pendulum. With the knowledge we already have of the changes of the field, how shall we describe everything that is going on here, in the field language?

The oscillation of the charge produces a changing electric field. This is always accompanied by a changing magnetic field. If a wire forming a closed circuit is placed in the vicinity, then again the changing magnetic field will be accompanied by an electric current in the circuit. All this is merely a repetition of known facts, but the study of Maxwell's equations gives a much deeper insight into the problem of the oscillating electric charge. By mathematical deduction from Maxwell's equations we can detect the character of the field surrounding an oscillating charge, its structure near and far from the source and its change with time. The outcome of such deduction is the *electromagnetic wave*. Energy radiates from the oscillating charge traveling with a definite speed through space; but a transference of energy, the motion of a state, is characteristic of all wave phenomena.

Different types of waves have already been considered. There was the longitudinal wave caused by the pulsating sphere, where the changes of density were propagated through the medium. There was the jelly-like medium in which the transverse wave spread. A deformation of the jelly, caused by the rotation of the sphere, moved through the medium. What kind of changes are now spreading in the case of an electromagnetic wave? Just the changes of an electromagnetic field! Every change of an electric field produces a magnetic field; every change of this magnetic field produces an electric field; every change of . . . , and so on. As field represents energy, all these changes spreading out in space, with a definite velocity, produce a wave. The electric and magnetic lines of force always lie, as deduced from the theory, on planes perpendicular to the direction of propagation. The wave produced is, therefore, transverse. The original features of the picture of the field we formed from Oersted's and Faraday's experiments are still preserved, but we now recognize that it has a deeper meaning.

The electromagnetic wave spreads in empty space.

This, again, is a consequence of the theory. If the oscillating charge suddenly ceases to move, then, its field becomes electrostatic. But the series of waves created by the oscillation continues to spread. The waves lead an independent existence and the history of their changes can be followed just as that of any other material object.

We understand that our picture of an electromagnetic wave, spreading with a certain velocity in space and changing in time, follows from Maxwell's equations only because they describe the structure of the electromagnetic field at any point in space and for any instant.

There is another very important question. With what speed does the electromagnetic wave spread in empty space? The theory, with the support of some data from simple experiments having nothing to do with the actual propagation of waves, gives a clear answer: *the velocity of an electromagnetic wave is equal to the velocity of light.*

Oersted's and Faraday's experiments formed the basis on which Maxwell's laws were built. All our results so far have come from a careful study of these laws, expressed in the field language. The theoretical discovery of an electromagnetic wave spreading with the speed of light is one of the greatest achievements in the history of science.

Experiment has confirmed the prediction of theory. Fifty years ago, Hertz proved, for the first time, the existence of electromagnetic waves and confirmed experimentally that their velocity is equal to that of light. Nowadays, millions of people demonstrate that electromagnetic waves are sent and received. Their apparatus is far more complicated than that used by Hertz and detects the presence of waves thousands of miles from their sources instead of only a few yards.

FIELD AND ETHER

The electromagnetic wave is a transverse one and is propagated with the velocity of light in empty space. The fact that their velocities are the same suggests a close relationship between optical and electromagnetic phenomena.

When we had to choose between the corpuscular and the wave theory, we decided in favor of the wave theory. The diffraction of light was the strongest argument influencing our decision. But we shall not contradict any of the explanations of the optical facts by also assuming that the *light wave is an electromagnetic one*. On the contrary, still other conclusions can be drawn. If this is really so, then there must exist some connection between the optical and electrical properties of matter that can be deduced from the theory. The fact that conclusions of this kind can really be drawn and that they stand the test of experiment is an essential argument in favor of the electromagnetic theory of light.

This great result is due to the field theory. Two apparently unrelated branches of science are covered by the same theory. The same Maxwell's equations describe both electric induction and optical refraction. If it is our aim to describe everything that ever happened or may happen with the help of one theory, then the union of optics and electricity is, undoubtedly, a very great step forward. From the physical point of view, the only difference between an ordinary electromagnetic wave and a light wave is the wave-length: this is very small for light waves, detected by the human eye, and great for ordinary electromagnetic waves, detected by a radio receiver.

The old mechanical view attempted to reduce all events in nature to forces acting between material particles. Upon this mechanical view was based the first naïve theory of the electric fluids. The field did not exist for the physicist of the early years of the nineteenth century. For him only substance and its changes were real. He tried to describe the action of two electric charges only by concepts referring directly to the two charges.

In the beginning, the field concept was no more than a means of facilitating the understanding of phenomena from the mechanical point of view. In the new field language it is the description of the field between the two charges, and not the charges themselves, which is essential for an understanding of their action. The recognition of the new concepts grew steadily, until substance was overshadowed by the field. It was realized that something of great importance had happened in physics. A new reality was created, a new concept for which there was no place in the mechanical description. Slowly and by a struggle the field concept established for itself a leading place in physics and has remained one of the basic physical concepts. The electromagnetic field is, for the modern physicist, as real as the chair on which he sits.

But it would be unjust to consider that the new field view freed science from the errors of the old theory of electric fluids or that the new theory destroys the achievements of the old. The new theory shows the merits as well as the limitations of the old theory and allows us to regain our old concepts from a higher level. This is true not only for the theories of electric fluids and field, but for all changes in physical theories, however revolutionary they may seem. In our case, we still find, for example, the concept of the electric charge in Maxwell's theory, though the charge is understood only as a source of the electric field. Coulomb's law is still valid and is contained in Maxwell's equations from which it can be deduced as one of the many consequences. We can still apply the old theory, whenever facts within the region of its validity are investigated. But we may as well apply the new theory, since all the known facts are contained in the realm of its validity.

To use a comparison, we could say that creating a new theory is not like destroying an old barn and erecting a skyscraper in its place. It is rather like climbing a mountain, gaining new and wider views, discovering unexpected connections between our starting point and its rich environment. But the point from which we started out still exists and can be seen, although it appears smaller and forms a tiny part of our broad view gained by the mastery of the obstacles on our adventurous way up.

It was, indeed, a long time before the full content of Maxwell's theory was recognized. The field was at first considered as something which might later be interpreted mechanically with the help of ether. By the time it was realized that this program could not be carried out, the achievements of the field theory had already become too striking and important for it to be exchanged for a mechanical dogma. On the other hand, the problem of devising the mechanical model of ether seemed to become less and less interesting and the result, in view of the forced and artificial character of the assumptions, more and more discouraging.

Our only way out seems to be to take for granted the fact that space has the physical property of transmitting electromagnetic waves, and not to bother too much about the meaning of this statement. We may still use the word ether, but only to express some physical property of space. This word ether has changed its meaning many times in the development of science. At the moment it no longer stands for a medium built up of particles. Its story, by no means finished, is continued by the relativity theory.

THE MECHANICAL SCAFFOLD

On reaching this stage of our story we must turn back to the beginning, to Galileo's law of inertia. We quote once more:

Every body perseveres in its state of rest, or of uniform motion in a right line, unless it is compelled to change that state by forces impressed thereon.

Once the idea of inertia is understood, one wonders what more can be said about it. Although this problem has already been thoroughly discussed, it is by no means exhausted.

Imagine a serious scientist who believes that the law of inertia can be proved or disproved by actual experiments. He pushes small spheres along a horizontal table, trying to eliminate friction so far as possible. He notices that the motion becomes more uniform as the table and the spheres are made smoother. Just as he is about to proclaim the principle of inertia, someone suddenly plays a practical joke on him. Our physicist works in a room without windows and has no communication whatever with the outside world. The practical joker installs some mechanism which enables him to cause the entire room to rotate quickly on an axis passing through its center. As soon as the rotation begins, the physicist has new and unexpected experiences. The sphere which has been moving uniformly tries to get as far away from the center and as near to the walls of the room as possible. He himself feels a strange force pushing him against the wall. He experiences the same sensation as anyone in a train or car traveling fast around a curve, or even more, on a rotating merry-go-round. All his previous results go to pieces.

Our physicist would have to discard, with the law of inertia, all mechanical laws. The law of inertia was his starting point; if this is changed so are all his further conclusions. An observer destined to spend his whole life in the rotating room and to perform all his experiments there, would have laws of mechanics differing from ours. If, on the other hand, he enters the room with a profound knowledge and a firm belief in the principles of physics, his explanation for the apparent breakdown of mechanics would be the assumption that the room rotates. By mechanical experiments he could even ascertain how it rotates.

Why should we take so much interest in the observer in his rotating room? Simply because we, on our earth, are to a certain extent, in the same position. Since the time of Copernicus we have known that the earth rotates on its axis and moves around the sun. Even this simple idea, so clear to everyone, was not left untouched by the advance of science. But let us leave this question for the time being and accept Copernicus' point of view. If our rotating observer could not confirm the laws of mechanics we, on our earth, should also be unable to do so. But the rotation of the earth is comparatively slow, so that the effect is not very distinct. Nevertheless there are many experiments which show a small deviation from the mechanical laws, and their consistency can be regarded as proof of the rotation of the earth.

Unfortunately we cannot place ourselves between the sun and the earth, to prove there the exact validity of the law of inertia and to get a view of the rotating earth. This can be done only in imagination. All our experiments must be performed on the earth on which we are compelled to live. The same fact is often expressed more scientifically: *the earth is our co-ordinate system.*

To show the meaning of these words more clearly, let us take a simple example. We can predict the position, at any time, of a stone thrown from a tower, and confirm our prediction by observation. If a measuring-rod is placed beside the tower, we can foretell with what point of the rod the falling body will coincide at any given moment. The tower and scale must, obviously, not be made of rubber or any other material which would undergo any change during the experiment. In fact, the unchangeable scale, rigidly connected with the earth, and a good clock are all we need, in principle, for the experiment. If we have these, we can ignore not only the architecture of the tower, but its very presence. The foregoing assumptions are all trivial and not usually specified in descriptions of such experiments. But this analysis shows how many hidden assumptions there are in every one of our statements. In our case, we assumed the existence of a rigid bar and an ideal clock, without which it would be impossible to check Galileo's law for falling bodies. With this simple but fundamental physical apparatus, a rod and a clock, we can confirm this mechanical law with a certain degree of accuracy. Carefully performed, this experiment reveals discrepancies between theory and experiment due to the rotation of the earth or, in other words, to the fact that the laws of mechanics, as here formulated, are not strictly valid in a co-ordinate system rigidly connected with the earth.

In all mechanical experiments, no matter of what type, we have to determine positions of material points at some definite time, just as in the above experiment with a falling body. But the position must always be described with respect to something, as in the previous case to the tower and the scale. We must have what we call some *frame of reference*, a mechanical scaffold, to be able to determine the positions of bodies. In describing the positions of objects and men in a city, the streets and avenues form the frame to which we refer. So far we have not bothered to describe the frame when quoting the laws of mechanics, because we happen to live on the earth and there is no difficulty in any particular case in fixing a frame of reference, rigidly connected with the earth. This frame, to which we refer all our observations, constructed of rigid unchangeable bodies, is called the *co-ordinate system*. Since this expression will be used very often, we shall simply write CS.

All our physical statements thus far have lacked something. We took no notice of the fact that all observations must be made in a certain CS. Instead of describing the structure of this CS we just ignored its existence. For example, when we wrote "a body moves uniformly . . ." we should really have written, "A body moves uniformly, relative to a chosen CS. . . ." Our experience with the rotating room taught us that the results of mechanical experiments may depend on the CS chosen.

If two CS rotate with respect to each other, then the laws of mechanics cannot be valid in both. If the surface of the water in a swimming pool, forming one of the co-ordinate systems, is horizontal, then in the other the surface of the water in a similar swimming pool takes the curved form familiar to anyone who stirs his coffee with a spoon.

When formulating the principal clews of mechanics we omitted one important point. We did not state for which CS they are valid. For this reason, the whole of classical mechanics hangs in mid-air since we do not know to which frame it refers. Let us, however, pass over this difficulty for the moment. We shall make the slightly incorrect assumption that in every CS rigidly connected with the earth, the laws of classical mechanics are valid. This is done in order to fix the CS and to make our statements definite. Although our statement that the earth is a suitable frame of reference is not wholly correct, we shall accept it for the present.

We assume, therefore, the existence of one CS for which the laws of mechanics are valid. Is this the only one? Suppose we have a CS such as a train, a ship or an airplane moving relative to our earth. Will the laws of mechanics be valid for these new CS? We know definitely that they are not always valid as for instance in the case of a train turning a curve, a ship tossed in a storm, or an airplane in a tail spin. Let us begin with a simple example. A CS moves uniformly, relative to our "good" CS, that is, one in which the laws of mechanics are valid. For instance, an ideal train or a ship sailing with delightful smoothness along a straight line and with a never-changing speed. We know from everyday experience that both systems will be "good," that physical experiments performed in a uniformly moving train or ship will give exactly the same results as on the earth. But, if the train stops, or accelerates abruptly, or if the sea is rough, strange things happen. In the train, the trunks fall off the luggage racks, on the ship tables and chairs are thrown about and the passengers become seasick. From the physical point of view this simply means that the laws of mechanics cannot be applied to these CS, that they are "bad" CS.

This result can be expressed by the so-called *Galilean relativity principle*: *if the laws of mechanics are valid in one CS, then they are valid in any other CS moving uniformly relative to the first.*

If we have two CS moving non-uniformly, relative to each other, then the laws of mechanics cannot be valid in both. "Good" co-ordinate systems, that is, those for which the laws of mechanics are valid, we call *inertial systems*. The question as to whether an inertial system exists at all is still unsettled. But if there is one such system, then there is an infinite number of them. Every CS moving uniformly, relative to the initial one, is also an inertial CS.

Let us consider the case of two CS starting from a
known position and moving uniformly, one relative
to the other, with a known velocity. One who prefers
concrete pictures can safely think of a ship or a train
moving relative to the earth. The laws of mechanics
can be confirmed experimentally with the same degree
of accuracy, on the earth or in a train or on a ship moving uniformly. But some difficulty arises if the observers of two systems begin to discuss observations of the
same event from the point of view of their different
CS. Each would like to translate the other's observations into his own language. Again a simple example:
the same motion of a particle is observed from two CS,
the earth and a train moving uniformly. These are
both inertial. Is it sufficient to know what is observed
in one CS in order to find out what is observed in the
other, if the relative velocities and positions of the two
CS at some moment are known? It is most essential, for
a description of events, to know how to pass from one
CS to another, since both CS are equivalent and both
equally suited for the description of events in nature.
Indeed, it is quite enough to know the results obtained
by an observer in one CS to know those obtained by an
observer in the other.

Let us consider the problem more abstractly, without ship or train. To simplify matters we shall investigate only motion along straight lines. We have, then,
a rigid bar with a scale and a good clock. The rigid
bar represents, in the simple case of rectilinear motion,
a CS just as did the scale on the tower in Galileo's experiment. It is always simpler and better to think of a
CS as a rigid bar in the case of rectilinear motion and
a rigid scaffold built of parallel and perpendicular rods

in the case of arbitrary motion in space, disregarding towers, walls, streets, and the like. Suppose we have, in our simplest case, two CS, that is two rigid rods; we draw one above the other and call them respectively the "upper" and "lower" CS. We assume that the two CS move with a definite velocity relative to each other, so that one slides along the other. It is safe to assume that both rods are of infinite length and have initial points but no end points. One clock is sufficient for the two CS, for the time flow is the same for both. When we begin our observation the starting points of the two rods coincide. The position of a material point is characterized, at this moment, by the same number in both CS. The material point coincides with a point on the scale on the rod, thus giving us a number determining the position of this material point. But, if the rods move uniformly, relative to each other, the numbers corresponding to the positions will be different after some time, say, one second. Consider a material point resting on the upper rod. The number determining its position on the upper CS does not change with time. But the corresponding number for the lower rod will change. Instead of "the number corresponding to a position of the point," we shall say briefly, the *co-*

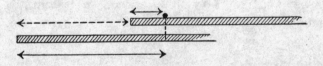

ordinate of a point. Thus we see from our drawing that although the following sentence sounds intricate, it is nevertheless correct and expresses something very simple. The co-ordinate of a point in the lower CS is equal to its co-ordinate in the upper CS plus the co-ordinate of the origin of the upper CS relative to the lower CS. The important thing is that we can always calculate the position of a particle in one CS if we know the position in the other. For this purpose we have to know the relative positions of the two co-ordinate systems in question at every moment. Although all this sounds learned it is, really, very simple and hardly worth such detailed discussion, except that we shall find it useful later.

It is worth our while to notice the difference between determining the position of a point and the time of an event. Every observer has his own rod which forms his CS, but there is only one clock for them all. Time is something "absolute" which flows in the same way for all observers in all CS.

Now another example. A man strolls with a velocity of three miles per hour along the deck of a large ship. This is his velocity relative to the ship, or, in other words, relative to a CS rigidly connected with the ship. If the velocity of the ship is thirty miles per hour relative to the shore, and if the uniform velocities of man and ship both have the same direction, then the velocity of the stroller will be thirty-three miles per hour

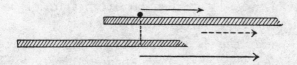

relative to an observer on the shore, or three miles per hour relative to the ship. We can formulate this fact more abstractly: the velocity of a moving material point, relative to the lower CS, is equal to that relative to the upper CS plus or minus the velocity of the upper CS relative to the lower, depending upon whether the velocities have the same or opposite directions. We can, therefore, always transform not only positions, but also velocities, from one CS to another if we know the relative velocities of the two CS. The positions, or co-ordinates, and velocities are examples of quantities which are different in different CS bound together by certain, in this case very simple, *transformation laws*.

There exist quantities, however, which are the same in both CS and for which no transformation laws are needed. Take as an example not one, but two fixed points on the upper rod and consider the distance between them. This distance is the difference in the co-ordinates of the two points. To find the positions of two points relative to different CS, we have to use transformation laws. But in constructing the differ-

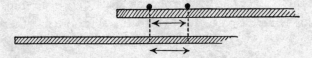

ences of two positions the contributions due to the different CS cancel each other and disappear, as is evident from the drawing. We have to add and subtract the distance between the origins of two CS. The distance of two points is, therefore, *invariant*, that is, independent of the choice of the CS.

The next example of a quantity independent of the CS is the change of velocity, a concept familiar to us from mechanics. Again, a material point moving along a straight line is observed from two CS. Its change of velocity is, for the observer in each CS, a difference between two velocities, and the contribution due to the uniform relative motion of the two CS disappears when the difference is calculated. Therefore, the change of velocity is an invariant, though only, of course, on condition that the relative motion of our two CS is uniform. Otherwise, the change of velocity would be different in each of the two CS, the difference being brought in by the change of velocity of the relative motion of the two rods, representing our co-ordinate systems.

Now the last example! We have two material points, with forces acting between them which depend only on the distance. In the case of rectilinear motion, the distance, and therefore the force as well, is invariant. Newton's law connecting the force with the change of velocity will, therefore, be valid in both CS. Once again we reach a conclusion which is confirmed by everyday experience: if the laws of mechanics are valid in one CS, then they are valid in all CS moving uniformly with respect to that one. Our example was, of course, a very simple one, that of rectilinear motion in which the CS can be represented by a rigid rod. But our conclusions are generally valid, and can be summarized as follows:

1. We know of no rule for finding an inertial system. Given one, however, we can find an infinite number, since all CS moving uniformly, relative to each other, are inertial systems if one of them is.

2. The time corresponding to an event is the same in all CS. But the co-ordinates and velocities are different, and change according to the transformation laws.

3. Although the co-ordinates and velocity change

when passing from one CS to another, the force and change of velocity, and therefore the laws of mechanics are invariant with respect to the transformation laws.

The laws of transformation formulated here for coordinates and velocities we shall call the transformation laws of classical mechanics, or more briefly, the *classical transformation*.

ETHER AND MOTION

The Galilean relativity principle is valid for mechanical phenomena. The same laws of mechanics apply to all inertial systems moving relative to each other. Is this principle also valid for nonmechanical phenomena, especially for those for which the field concepts proved so very important? All problems concentrated around this question immediately bring us to the starting point of the relativity theory.

We remember that the velocity of light *in vacuo*, or in other words, in ether, is 186,000 miles per second and that light is an electromagnetic wave spreading through the ether. The electromagnetic field carries energy which, once emitted from its source, leads an independent existence. For the time being, we shall continue to believe that the ether is a medium through which electromagnetic waves, and thus also light waves, are propagated, even though we are fully aware of the many difficulties connected with its mechanical structure.

We are sitting in a closed room so isolated from the external world that no air can enter or escape. If we sit still and talk we are, from the physical point of view, creating sound waves, which spread from their resting source with the velocity of sound in air. If there were no air or other material medium between the mouth and the ear, we could not detect a sound. Experiment has shown that the velocity of sound in air is the same in all directions, if there is no wind and the air is at rest in the chosen CS.

Let us now imagine that our room moves uniformly through space. A man outside sees, through the glass walls of the moving room (or train if you prefer) everything which is going on inside. From the measurements of the inside observer he can deduce the velocity of sound relative to his CS connected with his surroundings, relative to which the room moves. Here again is the old, much discussed, problem of determining the velocity in one CS if it is already known in another.

The observer in the room claims: the velocity of sound is, for me, the same in all directions.

The outside observer claims: the velocity of sound, spreading in the moving room and determined in my CS, is not the same in all directions. It is greater than the standard velocity of sound in the direction of the motion of the room and smaller in the opposite direction.

These conclusions are drawn from the classical transformation and can be confirmed by experiment. The room carries within it the material medium, the air through which sound waves are propagated, and the velocities of sound will, therefore, be different for the inside and outside observer.

We can draw some further conclusions from the

theory of sound as a wave propagated through a material medium. One way, though by no means the simplest, of not hearing what someone is saying, is to run, with a velocity greater than that of sound, relative to the air surrounding the speaker. The sound waves produced will then never be able to reach our ears. On the other hand, if we missed an important word which will never be repeated, we must run with a speed greater than that of sound to reach the produced wave and to catch the word. There is nothing irrational in either of these examples except that in both cases we should have to run with a speed of about four hundred yards per second, and we can very well imagine that further technical development will make such speeds possible. A bullet fired from a gun actually moves with a speed greater than that of sound and a man placed on such a bullet would never hear the sound of the shot.

All these examples are of a purely mechanical character and we can now formulate the important questions: could we repeat what has just been said of a sound wave, in the case of a light wave? Do the Galilean relativity principle and the classical transformation apply to mechanical as well as to optical and electrical phenomena? It would be risky to answer these questions by a simple "yes" or "no" without going more deeply into their meaning.

In the case of the sound wave in the room moving uniformly, relative to the outside observer, the following intermediate steps are very essential for our conclusion:

The moving room carries the air in which the sound
    wave is propagated.

The velocities observed in two CS moving uniformly, relative to each other, are connected by the classical transformation.

The corresponding problem for light must be formulated a little differently. The observers in the room are no longer talking, but are sending light signals, or light waves in every direction. Let us further assume that the sources emitting the light signals are permanently resting in the room. The light' waves move through the ether just as the sound waves moved through the air.

Is the ether carried with the room as the air was? Since we have no mechanical picture of the ether it is extremely difficult to answer this question. If the room is closed, the air inside is forced to move with it. There is obviously no sense in thinking of ether in this way, since all matter is immersed in it and it penetrates everywhere. No doors are closed to ether. The "moving room," now means only a moving CS to which the source of light is rigidly connected. It is, however, not beyond us to imagine that the room moving with its light source carries the ether along with it just as the sound source and air were carried along in the closed room. But we can equally well imagine the opposite: that the room travels through the ether as a ship through a perfectly smooth sea, not carrying any part of the medium along but moving through it. In our first picture, the room moving with its light source carries the ether. An analogy with a sound wave is possible and quite similar conclusions can be drawn. In the second, the room moving with its light source does not carry the ether. No analogy with a sound wave is possible and the conclusions drawn in the case of a sound wave do not hold for a light wave. These are the two limiting possibilities. We could imagine the still more complicated possibility that the ether is only partially carried by the room moving with its light source. But there is no reason to discuss the more complicated assumptions before finding out which of the two simpler limiting cases experiment favors.

We shall begin with our first picture and assume, for the present: the ether is carried along by the room moving with its rigidly-connected light source. If we believe in the simple transformation principle for the velocities of sound waves, we can now apply our conclusions to light waves as well. There is no reason for doubting the simple mechanical transformation law which only states that the velocities have to be added in certain cases and subtracted in others. For the moment, therefore, we shall assume both the carrying of the ether by the room moving with its light source and the classical transformation.

If I turn on the light and its source is rigidly connected with my room, then the velocity of the light signal has the well-known experimental value 186,000 miles per second. But the outside observer will notice the motion of the room, and, therefore, that of the source and, since the ether is carried along, his conclusion must be: the velocity of light in my outside CS is different in different directions. It is greater than the standard velocity of light in the direction of the motion of the room and smaller in the opposite direction. Our conclusion is: if ether is carried with the room moving with its light source and if the mechanical laws are valid, then the velocity of light must depend on the velocity of the light source. Light reaching our eyes from a moving light source would have a greater velocity if the motion is toward us and smaller if it is away from us.

If our speed were greater than that of light we should be able to run away from a light signal. We could see occurrences from the past by reaching previously sent light waves. We should catch them in a reverse order to that in which they were sent, and the train of happenings on our earth would appear like a film shown backward, beginning with a happy ending. These conclusions all follow from the assumption that the moving CS carries along the ether and the mechanical transformation laws are valid. If this is so, the analogy between light and sound is perfect.

But there is no indication as to the truth of these conclusions. On the contrary, they are contradicted by all observations made with the intention of proving them. There is not the slightest doubt as to the clarity of this verdict, although it is obtained through rather indirect experiments in view of the great technical difficulties caused by the enormous value of the velocity of light. *The velocity of light is always the same in all CS independent of whether or not the emitting source moves, or how it moves.*

We shall not go into detailed description of the many experiments from which this important conclusion can be drawn. We can, however, use some very simple arguments which, though they do not prove that the velocity of light is independent of the motion of the source, nevertheless make this fact convincing and understandable.

In our planetary system the earth and other planets move around the sun. We do not know of the existence of other planetary systems, similar to ours. There are, however, very many double-star systems, consisting of two stars moving around a point, called their center of gravity. Observation of the motion of these double stars reveals the validity of Newton's gravitational law. Now suppose that the speed of light depends on the velocity of the emitting body. Then the message, that is, the light ray from the star, will travel more quickly or more slowly, according to the velocity of the star at the moment the ray is emitted. In this case the whole motion would be muddled and it would be impossible to confirm, in the case of distant double stars, the validity of the same gravitational law which rules over our planetary system.

Let us consider another experiment based upon a very simple idea. Imagine a wheel rotating very quickly. According to our assumption, the ether is carried by the motion and takes a part in it. A light wave passing near the wheel would have a different speed when the wheel is at rest than when it is in motion. The velocity of light in ether at rest should differ from that in ether which is being quickly dragged round by the motion of the wheel, just as the velocity of a sound wave varies on calm and windy days. But no such difference is detected! No matter from which angle we approach the subject, or what crucial experiment we may devise, the verdict is always against the assumption of the ether carried by motion. Thus, the result of our considerations, supported by more detailed and technical argument, is:

The velocity of light does not depend on the motion
of the emitting source.

It must not be assumed that the moving body carries
the surrounding ether along.

We must, therefore, give up the analogy between
sound and light waves and turn to the second possibility: that all matter moves through the ether, which
takes no part whatever in the motion. This means that
we assume the existence of a sea of ether with all CS
resting in it, or moving relative to it. Suppose we
leave, for a while, the question as to whether experiment proved or disproved this theory. It will be better
to become more familiar with the meaning of this new
assumption and with the conclusions which can be
drawn from it.

There exists a CS resting relative to the ether-sea.
In mechanics, not one of the many CS moving uniformly, relative to each other, could be distinguished.
All such CS were equally "good" or "bad." If we have
two CS moving uniformly, relative to each other, it is
meaningless, in mechanics, to ask which of them is in
motion and which at rest. Only relative uniform motion can be observed. We cannot talk about absolute
uniform motion because of the Galilean relativity
principle. What is meant by the statement that *absolute*
and not only *relative* uniform motion exists? Simply
that there exists one CS in which some of the laws of
nature are different from those in all others. Also that
every observer can detect whether his CS is at rest or
in motion by comparing the laws valid in it with those
valid in the only one which has the absolute monopoly
of serving as the standard CS. Here is a different state

of affairs from classical mechanics, where absolute uniform motion is quite meaningless because of Galileo's law of inertia.

What conclusions can be drawn in the domain of field phenomena if motion through ether is assumed? This would mean that there exists one CS distinct from all others, at rest relative to the ether-sea. It is quite clear that some of the laws of nature must be different in this CS, otherwise the phrase, "motion through ether," would be meaningless. If the Galilean relativity principle is valid then motion through ether makes no sense at all. It is impossible to reconcile these two ideas. If, however, there exists one special CS fixed by the ether, then to speak of "absolute motion" or "absolute rest," has a definite meaning.

We really have no choice. We tried to save the Galilean relativity principle by assuming that systems carry the ether along in their motion, but this led to a contradiction with experiment. The only way out is to abandon the Galilean relativity principle and try out the assumption that all bodies move through the calm ether-sea.

The next step is to consider some conclusions contradicting the Galilean relativity principle and supporting the view of motion through ether, and to put them to the test of an experiment. Such experiments are easy enough to imagine, but very difficult to perform. As we are concerned here only with ideas, we need not bother with technical difficulties.

Again we return to our moving room with two observers, one inside and one outside. The outside observer will represent the standard CS, designated by the ether-sea. It is the distinguished CS in which the velocity of light always has the same standard value. All light sources, whether moving or at rest in the calm ether-sea, propagate light with the same velocity. The room and its observer move through the ether. Imagine that a light in the center of the room is flashed on and off and, furthermore, that the walls of the room are transparent so that the observers, both inside and outside, can measure the velocity of the light. If we ask the two observers what results they expect to obtain, their answers would run something like this:

The outside observer: My CS is designated by the ether-sea. Light in my CS always has the standard value. I need not care whether or not the source of light or other bodies are moving, for they never carry my ether-sea with them. My CS is distinguished from all others and the velocity of light must have its standard value in this CS, independent of the direction of the light beam or the motion of its source.

The inside observer: My room moves through the ether-sea. One of the walls runs away from the light and the other approaches it. If my room traveled, relative to the ether-sea, with the velocity of light, then the light emitted from the center of the room would never reach the wall running away with the velocity of light. If the room traveled with a velocity smaller than that of light, then a wave sent from the center of the room would reach one of the walls before the other. The wall moving toward the light wave would be reached before the one retreating from the light wave. Therefore, although the source of light is rigidly connected with my CS, the velocity of light will not be the same in all directions. It will be smaller in the direction of the motion relative to the ether-sea as the wall runs away, and greater in the opposite direction as the wall moves toward the wave and tries to meet it sooner.

Thus, only in the one CS distinguished by the ether-sea should the velocity of light be equal in all directions. For other CS moving relatively to the ether-sea it should depend on the direction in which we are measuring.

The crucial experiment just considered enables us to test the theory of motion through the ether-sea. Nature, in fact, places at our disposal a system moving with a fairly high velocity: the earth in its yearly motion around the sun. If our assumption is correct, then the velocity of light in the direction of the motion of the earth should differ from the velocity of light in an opposite direction. The differences can be calculated and a suitable experimental test devised. In view of the small time-differences following from the theory, very ingenious experimental arrangements have to be thought out. This was done in the famous Michelson-Morley experiment. The result was a verdict of "death" to the theory of a calm ether-sea through which all matter moves. No dependence of the speed of light upon direction could be found. Not only the speed of light, but also other field phenomena would show a dependence on the direction in the moving CS, if the theory of the ether-sea were assumed. Every experiment has given the same negative result as the Michelson-Morley one, and never revealed any dependence upon the direction of the motion of the earth.

The situation grows more and more serious. Two assumptions have been tried. The first, that moving bodies carry ether along. The fact that the velocity of light does not depend on the motion of the source contradicts this assumption. The second, that there exists one distinguished CS and that moving bodies do not carry the ether but travel through an ever calm ether-sea. If this is so, then the Galilean relativity principle is not valid and the speed of light cannot be the same in every CS. Again we are in contradiction with experiment.

More artificial theories have been tried out, assuming that the real truth lies somewhere between these two limiting cases: that the ether is only partially carried by the moving bodies. But they all failed! Every attempt to explain the electromagnetic phenomena in moving CS with the help of the motion of the ether, motion through the ether, or both these motions, proved unsuccessful.

Thus arose one of the most dramatic situations in the history of science. All assumptions concerning ether led nowhere! The experimental verdict was always negative. Looking back over the development of physics we see that the ether, soon after its birth, became the *"enfant terrible"* of the family of physical substances. First, the construction of a simple mechanical picture of the ether proved to be impossible and was discarded. This caused, to a great extent, the breakdown of the mechanical point of view. Second, we had to give up hope that through the presence of the ether-sea one CS would be distinguished and lead to the recognition of absolute, and not only relative, motion. This would have been the only way, besides carrying the waves, in which ether could mark and justify its existence. All our attempts to make ether real failed. It revealed neither its mechanical construction nor absolute motion. Nothing remained of all the properties of the ether except that for which it was invented, i.e., its ability to transmit electromagnetic waves. Our attempts to discover the properties of the ether led to difficulties and contradictions. After such bad experiences, this is the moment to forget the ether completely and to try never to mention its name. We shall say: our space has the physical property of transmitting waves, and so omit the use of a word we have decided to avoid.

The omission of a word from our vocabulary is, of course, no remedy. Our troubles are indeed much too profound to be solved in this way!

Let us now write down the facts which have been sufficiently confirmed by experiment without bothering any more about the "e——r" problem.

1. The velocity of light in empty space always has its standard value, independent of the motion of the source or receiver of light.
2. In two CS moving uniformly, relative to each other, all laws of nature are exactly identical and there is no way of distinguishing absolute uniform motion.

There are many experiments to confirm these two statements and not a single one to contradict either of them. The first statement expresses the constant character of the velocity of light, the second generalizes the Galilean relativity principle, formulated for mechanical phenomena, to all happenings in nature.

In mechanics, we have seen: If the velocity of a material point is so and so, relative to one CS, then it will be different in another CS moving uniformly, relative to the first. This follows from the simple mechanical transformation principles. They are immediately given by our intuition (man moving relative to ship and shore) and apparently nothing can be wrong here! But this transformation law is in contradiction to the constant character of the velocity of light. Or, in other words, we add a third principle:

3. Positions and velocities are transformed from one inertial system to another according to the classical transformation.

   The contradiction is then evident. We cannot combine (1), (2), and (3).

The classical transformation seems too obvious and simple for any attempt to change it. We have already tried to change (1) and (2) and came to a disagreement with experiment. All theories concerning the motion of "e——r" required an alteration of (1) and (2). This was no good. Once more we realize the serious character of our difficulties. A new clew is needed. It is supplied by *accepting the fundamental assumptions (1) and (2)*, and, strange though it seems, *giving up (3)*. The new clew starts from an analysis of the most fundamental and primitive concepts; we shall show how this analysis forces us to change our old views and removes all our difficulties.

TIME, DISTANCE, RELATIVITY

Our new assumptions are:

1. *The velocity of light* in vacuo *is the same in all CS moving uniformly, relative to each other.*
2. *All laws of nature are the same in all CS moving uniformly, relative to each other.*

The *relativity theory* begins with these two assump-

tions. From now on we shall not use the classical transformation because we know that it contradicts our
assumptions.

It is essential here, as always in science, to rid ourselves of deep-rooted, often uncritically repeated, prejudices. Since we have seen that changes in (1) and (2)
lead to contradiction with experiment, we must have
the courage to state their validity clearly and to attack
the one possibly weak point, the way in which positions and velocities are transformed from one CS to
another. It is our intention to draw conclusions from
(1) and (2), see where and how these assumptions
contradict the classical transformation, and find the
physical meaning of the results obtained.

Once more, the example of the moving room with
outside and inside observers will be used. Again a light
signal is emitted from the center of the room and again
we ask the two men what they expect to observe, assuming only our two principles and forgetting what
was previously said concerning the medium through
which the light travels. We quote their answers:

The inside observer: The light signal traveling from
the center of the room will reach the walls *simultaneously*, since all the walls are equally distant from the
light source and the velocity of light is the same in all
directions.

The outside observer: In my system, the velocity of
light is exactly the same as in that of the observer moving with the room. It does not matter to me whether
or not the light source moves in my CS since its motion
does not influence the velocity of light. What I see is a
light signal traveling with a standard speed, the same
in all directions. One of the walls is trying to escape

from and the opposite wall to approach the light signal. Therefore, the escaping wall will be met by the signal a little later than the approaching one. Although the difference will be very slight if the velocity of the room is small compared with that of light, the light signal will nevertheless not meet these two opposite walls, which are perpendicular to the direction of the motion, quite simultaneously.

Comparing the predictions of our two observers we find a most astonishing result which flatly contradicts the apparently well-founded concepts of classical physics. Two events, i.e., the two light beams reaching the two walls, are simultaneous for the observer on the inside, but not for the observer on the outside. In classical physics, we had one clock, one time flow, for all observers in all CS. Time, and therefore such words as "simultaneously," "sooner," "later," had an absolute meaning independent of any CS. Two events happening at the same time in one CS happened necessarily simultaneously in all other CS.

Assumptions (1) and (2), i.e., the relativity theory, force us to give up this view. We have described two events happening at the same time in one CS, but at different times in another CS. Our task is to understand this consequence, to understand the meaning of the sentence: "Two events which are simultaneous in one CS, may not be simultaneous in another CS."

What do we mean by "two simultaneous events in one CS"? Intuitively everyone seems to know the meaning of this sentence. But let us make up our minds to be cautious and try to give rigorous definitions, as we know how dangerous it is to overestimate intuition. Let us first answer a simple question.

What is a clock?

The primitive subjective feeling of time flow enables us to order our impressions, to judge that one event takes place earlier, another later. But to show that the time interval between two events is 10 seconds, a clock is needed. By the use of a clock the time concept becomes objective. Any physical phenomenon may be used as a clock, provided it can be exactly repeated as many times as desired. Taking the interval between the beginning and the end of such an event as one unit of time, arbitrary time-intervals may be measured by repetition of this physical process. All clocks, from the simple hourglass to the most refined instruments, are based on this idea. With the hourglass the unit of time is the interval the sand takes to flow from the upper to the lower glass. The same physical process can be repeated by inverting the glass.

At two distant points we have two perfect clocks, showing exactly the same time. This statement should be true regardless of the care with which we verify it. But what does it really mean? How can we make sure that distant clocks always show exactly the same time? One possible method would be to use television. It should be understood that television is used only as an example and is not essential to our argument. I could stand near one of the clocks and look at a televised picture of the other. I could then judge whether or not they showed the same time simultaneously. But this would not be a good proof. The televised picture is transmitted through electromagnetic waves and thus travels with the speed of light. Through television I see a picture which was sent some very short time before, whereas on the real clock I see what is taking place at the present moment. This difficulty can easily be avoided. I must take television pictures of the two clocks at a point equally distant from each of them and observe them from this center point. Then, if the signals are sent out simultaneously, they will all reach me at the same instant. If two good clocks observed from the mid-point of the distance between them always show the same time, then they are well suited for designating the time of events at two distant points.

In mechanics we used only one clock. But this was not very convenient, because we had to take all measurements in the vicinity of this one clock. Looking at the clock from a distance, for example by television, we have always to remember that what we see now really happened earlier, just as we receive light from the sun eight minutes after it was emitted. We should have to make corrections, according to our distance from the clock, in all our time readings.

It is, therefore, inconvenient to have only one clock. Now, however, as we know how to judge whether two, or more, clocks show the same time simultaneously and run in the same way, we can very well imagine as many clocks as we like in a given CS. Each of them will help us to determine the time of the events happening in its immediate vicinity. The clocks are all at rest relative to the CS. They are "good" clocks and are *synchronized*, which means that they show the same time simultaneously.

There is nothing especially striking or strange about the arrangements of our clocks. We are now using many synchronized clocks instead of only one and can, therefore, easily judge whether or not two distant events are simultaneous in a given CS. They are if the synchronized clocks in their vicinity show the same time at the instant the events happen. To say that one of the distant events happens before the other has now a definite meaning. All this can be judged by the help of the synchronized clocks at rest in our CS.

This is in agreement with classical physics, and not one contradiction to the classical transformation has yet appeared.

For the definition of simultaneous events, the clocks are synchronized by the help of signals. It is essential in our arrangement that these signals travel with the velocity of light, the velocity which plays such a fundamental role in the theory of relativity.

Since we wish to deal with the important problem of two CS moving uniformly, relative to each other, we must consider two rods, each provided with clocks. The observer in each of the two CS moving relative to each other now has his own rod with his own set of clocks rigidly attached.

When discussing measurements in classical mechanics we used one clock for all CS. Here we have many clocks in each CS. This difference is unimportant. One clock was sufficient, but nobody could object to the use of many, so long as they behave as decent synchronized clocks should.

Now we are approaching the essential point showing where the classical transformation contradicts the theory of relativity. What happens when two sets of clocks are moving uniformly, relative to each other? The classical physicist would answer: Nothing; they still have the same rhythm, and we can use moving as well as resting clocks to indicate time. According to classical physics, two events simultaneous in one CS will also be simultaneous in any other CS.

But this is not the only possible answer. We can equally well imagine a moving clock having a different rhythm from one at rest. Let us now discuss this possibility without deciding, for the moment, whether or not clocks really change their rhythm in motion. What is meant by the statement that a moving clock changes its rhythm? Let us assume, for the sake of simplicity, that we have only one clock in the upper CS and many in the lower. All the clocks have the same mechanism, and the lower ones are synchronized, that is, they show the same time simultaneously. We have drawn three subsequent positions of the two CS moving relative to each other. In the first drawing the positions of the hands of the upper and lower clocks are, by convention, the same because we arranged them so. All the clocks show the same time. In the second drawing, we see the relative positions of the two CS some time later. All the clocks in the lower CS show the same time, but the clock in the upper CS is out of rhythm. The rhythm is changed and the time differs because the clock is moving relative to the lower CS. In the third drawing we see the difference in the positions of the hands increased with time.

An observer at rest in the lower CS would find that a moving clock changes its rhythm. Certainly the same result could be found if the clock moved relative to an observer at rest in the upper CS; in this case there would have to be many clocks in the upper CS and only one in the lower. The laws of nature must be the same in both CS moving relative to each other.

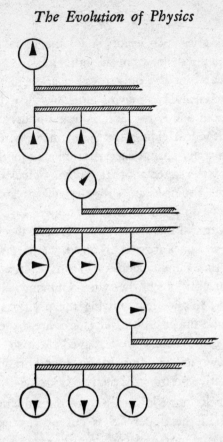

In classical mechanics it was tacitly assumed that a moving clock does not change its rhythm. This seemed so obvious that it was hardly worth mentioning. But nothing should be too obvious; if we wish to be really careful, we should analyze the assumptions, so far taken for granted, in physics.

An assumption should not be regarded as unreasonable simply because it differs from that of classical physics. We can well imagine that a moving clock changes its rhythm, so long as the law of this change is the same for all inertial CS.

Yet another example. Take a yardstick; this means that a stick is a yard in length as long as it is at rest in a CS. Now it moves uniformly, sliding along the rod representing the CS. Will its length still appear to be one yard? We must know beforehand how to determine its length. As long as the stick was at rest its ends coincided with markings one yard apart on the CS. From this we concluded: the length of the resting stick is one yard. How are we to measure this stick during motion? It could be done as follows. At a given moment two observers simultaneously take snapshots, one of the origin of the stick and the other of the end. Since the pictures are taken simultaneously we can compare the marks on the CS rod with which the origin and the end of the moving stick coincide. In this way we determine its length. There must be two observers to take note of simultaneous events in different parts of the given CS. There is no reason to believe that the result of such measurements will be the same as in the case of a stick at rest. Since the photographs had to be taken simultaneously, which is, as we already know, a relative concept depending on the CS, it seems quite possible that the results of this measurement will be different in different CS moving relative to each other.

We can well imagine that not only does the moving clock change its rhythm, but also that a moving stick changes its length, so long as the laws of the changes are the same for all inertial CS.

We have only been discussing some new possibilities without giving any justification for assuming them.

We remember: the velocity of light is the same in all inertial CS. It is impossible to reconcile this fact with the classical transformation. The circle must be broken somewhere. Can it not be done just here? Can we not assume such changes in the rhythm of the moving clock and in the length of the moving rod that the constancy of the velocity of light will follow directly from these assumptions? Indeed we can! Here is the first instance in which the relativity theory and classical physics differ radically. Our argument can be reversed: if the velocity of light is the same in all CS, then moving rods must change their length, moving clocks must change their rhythm, and the laws governing these changes are rigorously determined.

There is nothing mysterious or unreasonable in all this. In classical physics it was always assumed that clocks in motion and at rest have the same rhythm, that rods in motion and at rest have the same length. If the velocity of light is the same in all CS, if the relativity theory is valid, then we must sacrifice this assumption. It is difficult to get rid of deep-rooted prejudices, but there is no other way. From the point of view of the relativity theory the old concepts seem arbitrary. Why believe, as we did some pages ago, in absolute time flowing in the same way for all observers in all CS? Why believe in unchangeable distance? Time is determined by clocks, space co-ordinates by rods, and the result of their determination may depend on the behavior of these clocks and rods when in motion. There is no reason to believe that they will behave in the way we should like them to. Observation shows, indirectly, through the phenomena of electromagnetic field, that a moving clock changes its rhythm, a rod its length, whereas on the basis of mechanical phenomena we did not think this happened. We must accept the concept of relative time in every CS, because it is the best way out of our difficulties. Further scientific advance, developing from the theory of relativity, shows that this new aspect should not be regarded as a *malum necessarium*, for the merits of the theory are much too marked.

So far we have tried to show what led to the fundamental assumptions of the relativity theory, and how the theory forced us to revise and to change the classical transformation by treating time and space in a new way. Our aim is to indicate the ideas forming the basis of a new physical and philosophical view. These ideas are simple; but in the form in which they have been formulated here, they are insufficient for arriving at not only qualitative, but also quantitative conclusions. We must again use our old method of explaining only the principal ideas and stating some of the others without proof.

To make clear the difference between the view of the old physicist, whom we shall call O and who believes in the classical transformation, and that of the modern physicist, whom we shall call M and who knows the relativity theory, we shall imagine a dialogue between them.

O. I believe in the Galilean relativity principle in mechanics, because I know that the laws of mechanics are the same in two CS moving uniformly relative to each other, or in other words, that these laws are invariant with respect to the classical transformation.

M. But the relativity principle must apply to all events in our external world. Not only the laws of mechanics but all laws of nature must be the same in CS moving uniformly, relative to each other.

O. But how can all laws of nature possibly be the

same in CS moving relative to each other? The field
equations, that is, Maxwell's equations, are not invariant with respect to the classical transformation. This is
clearly shown by the example of the velocity of light.
According to the classical transformation, this velocity
should not be the same in two CS moving relative to
each other.

M. This merely shows that the classical transformation cannot be applied, that the connection between
two CS must be different; that we may not connect
co-ordinates and velocities as is done in these transformation laws. We have to substitute new laws and deduce them from the fundamental assumptions of the
theory of relativity. Let us not bother about the mathematical expression for this new transformation law,
and be satisfied that it is different from the classical.
We shall call it briefly the *Lorentz transformation*. It
can be shown that Maxwell's equations, that is, the
laws of field are invariant with respect to the Lorentz
transformation, just as the laws of mechanics are invariant with respect to the classical transformation.
Remember how it was in classical physics. We had
transformation laws for co-ordinates, transformation
laws for velocities, but the laws of mechanics were the
same for two CS moving uniformly, relative to each
other. We had transformation laws for space, but not
for time, because time was the same in all CS. Here,
however, in the relativity theory, it is different. We
have transformation laws different from the classical
for space, time, and velocity. But again the laws of
nature must be the same in all CS moving uniformly,
relative to each other. The laws of nature must be invariant, not, as before, with respect to the classical

transformation, but with respect to a new type of transformation, the so-called Lorentz transformation. In all inertial CS the same laws are valid and the transition from one CS to another is given by the Lorentz transformation.

O. I take your word for it, but it would interest me to know the difference between the classical and Lorentz transformations.

M. Your question is best answered in the following way. Quote some of the characteristic features of the classical transformation and I shall try to explain whether or not they are preserved in the Lorentz transformation, and if not, how they are changed.

O. If something happens at some point at some time in my CS, then the observer in another CS moving uniformly, relative to mine, assigns a different number to the position in which this event occurs, but of course the same time. We use the same clock in all our CS and it is immaterial whether or not the clock moves. Is this also true for you?

M. No, it is not. Every CS must be equipped with its own clocks at rest, since motion changes the rhythm. Two observers in two different CS will assign not only different numbers to the position, but also different numbers to the time at which this event happens.

O. This means that the time is no longer an invariant. In the classical transformation it is always the same time in all CS. In the Lorentz transformation it changes and somehow behaves like the co-ordinate in the old transformation. I wonder how it is with distance? According to classical mechanics a rigid rod preserves its length in motion or at rest. Is this also true now?

M. It is not. In fact, it follows from the Lorentz transformation that a moving stick contracts in the direction of the motion and the contraction increases if the speed increases. The faster a stick moves, the

shorter it appears. But this occurs only in the direction of the motion. You see in my drawing a moving rod which shrinks to half its length, when it moves with a velocity approaching *ca.* 90 per cent of the velocity of

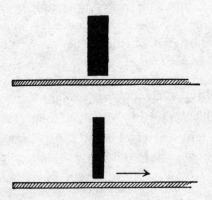

light. There is no contraction, however, in the direction perpendicular to the motion, as I have tried to illustrate in my last drawing.

O. This means that the rhythm of a moving clock and the length of a moving stick depend on the speed. But how?

M. The changes become more distinct as the speed increases. It follows from the Lorentz transformation that a stick would shrink to nothing if its speed were to reach that of light. Similarly the rhythm of a moving clock is slowed down, compared to the clocks it passes along the rod, and would come to a stop if the clock were to move with the speed of light, that is, if the clock is a "good" one.

O. This seems to contradict all our experience. We know that a car does not become shorter when in motion and we also know that the driver can always compare his "good" watch with those he passes on the way, finding that they agree fairly well, contrary to your statement.

M. This is certainly true. But these mechanical velocities are all very small compared to that of light, and it is, therefore, ridiculous to apply relativity to these phenomena. Every car driver can safely apply classical physics even if he increases his speed a hundred thousand times. We could only expect disagreement between experiment and the classical transformation with velocities approaching that of light. Only with very great velocities can the validity of the Lorentz transformation be tested.

O. But there is yet another difficulty. According to mechanics I can imagine bodies with velocities even greater than that of light. A body moving with the velocity of light relative to a floating ship moves with a velocity greater than that of light relative to the shore. What will happen to the stick which shrank to nothing when its velocity was that of light? We can hardly expect a negative length if the velocity is greater than that of light.

M. There is really no reason for such sarcasm! From the point of view of the relativity theory a material body cannot have a velocity greater than that of light. The velocity of light forms the upper limit of velocities for all material bodies. If the speed of a body is equal to that of light relative to a ship then it will also be equal to that of light relative to the shore. The simple mechanical law of adding and subtracting velocities is no longer valid or, more precisely, is only approximately valid for small velocities, but not for those near the velocity of light. The number expressing the velocity of light appears explicitly in the Lorentz transformation, and plays the role of a limiting case, like the infinite velocity in classical mechanics. This more general theory does not contradict the classical transformation and classical mechanics. On the contrary, we regain the old concepts as a limiting case when the velocities are small. From the point of view of the new theory it is clear in which cases classical physics is valid and wherein its limitations lie. It would be just as ridiculous to apply the theory of relativity to the motion of cars, ships, and trains as to use a calculating machine where a multiplication table would be sufficient.

RELATIVITY AND MECHANICS

The relativity theory arose from necessity, from serious and deep contradictions in the old theory from which there seemed no escape. The strength of the new theory lies in the consistency and simplicity with which it solves all these difficulties, using only a few very convincing assumptions.

Although the theory arose from the field problem, it

has to embrace all physical laws. A difficulty seems to appear here. The field laws on the one hand and the mechanical laws on the other are of quite different kinds. The equations of electromagnetic field are invariant with respect to the Lorentz transformation and the mechanical equations are invariant with respect to the classical transformation. But the relativity theory claims that all laws of nature must be invariant with respect to the Lorentz and not to the classical transformation. The latter is only a special, limiting case of the Lorentz transformation when the relative velocities of two CS are very small. If this is so, classical mechanics must change in order to conform with the demand of invariance with respect to the Lorentz transformation. Or, in other words, classical mechanics cannot be valid if the velocities approach that of light. Only one transformation from one CS to another can exist, namely, the Lorentz transformation.

It was simple to change classical mechanics in such a way that it contradicted neither the relativity theory nor the wealth of material obtained by observation and explained by classical mechanics. The old mechanics is valid for small velocities and forms the limiting case of the new one.

It would be interesting to consider some instance of a change in classical mechanics introduced by the relativity theory. This might, perhaps, lead us to some conclusions which could be proved or disproved by experiment.

Let us assume a body, having a definite mass, moving along a straight line, and acted upon by an external force in the direction of the motion. The force, as we know, is proportional to the change of velocity. Or, to be more explicit, it does not matter whether a given body increases its velocity in one second from 100 to 101 feet per second, or from 100 miles to 100 miles and one foot per second or from 180,000 miles to 180,000 miles and one foot per second. The force acting upon a particular body is always the same for the same change of velocity in the same time.

Is this sentence true from the point of view of the relativity theory? By no means! This law is valid only for small velocities. What, according to the relativity theory, is the law for great velocities, approaching that of light? If the velocity is great, extremely strong forces are required to increase it. It is not at all the same thing to increase by one foot per second a velocity of about 100 feet per second or a velocity approaching that of light. The nearer a velocity is to that of light the more difficult it is to increase. When a velocity is equal to that of light it is impossible to increase it further. Thus, the changes brought about by the relativity theory are not surprising. The velocity of light is the upper limit for all velocities. No finite force, no matter how great, can cause an increase in speed beyond this limit. In place of the old mechanical law connecting force and change of velocity, a more complicated one appears. From our new point of view classical mechanics is simple because in nearly all observations we deal with velocities much smaller than that of light.

A body at rest has a definite mass, called the *rest mass*. We know from mechanics that every body resists a change in its motion; the greater the mass, the stronger the resistance, and the weaker the mass, the weaker the resistance. But in the relativity theory, we have something more. Not only does a body resist a change more strongly if the rest mass is greater, but also if its velocity is greater. Bodies with velocities approaching that of light would offer a very strong resistance to external forces. In classical mechanics the resistance of a given body was something unchangeable, characterized by its mass alone. In the relativity theory it depends on both rest mass and velocity. The resistance becomes infinitely great as the velocity approaches that of light.

The results just quoted enable us to put the theory to the test of experiment. Do projectiles with a velocity approaching that of light resist the action of an external force as predicted by the theory? Since the statements of the relativity theory have, in this respect, a quantitative character, we could confirm or disprove the theory if we could realize projectiles having a speed approaching that of light.

Indeed, we find in nature projectiles with such velocities. Atoms of radioactive matter, radium for instance, act as batteries which fire projectiles with enormous velocities. Without going into detail we can quote only one of the very important views of modern physics and chemistry. All matter in the universe is made up of *elementary particles* of only a few kinds. It is like seeing in one town buildings of different sizes, construction and architecture, but from shack to skyscraper only very few different kinds of bricks were used, the same in all the buildings. So all known elements of our material world, from hydrogen the lightest, to uranium the heaviest, are built of the same kinds of bricks, that is, the same kinds of elementary particles. The heaviest elements, the most complicated buildings, are unstable and they disintegrate or, as we say, are *radioactive*. Some of the bricks, that is, the elementary particles of which the radioactive atoms are constructed, are sometimes thrown out with a very great velocity, approaching that of light. An atom of an element, say radium, according to our present views, confirmed by numerous experiments, is a complicated structure, and radioactive disintegration is one of those phenomena in which the composition of atoms from still simpler bricks, the elementary particles, is revealed.

By very ingenious and intricate experiments we can find out how the particles resist the action of an external force. The experiments show that the resistance offered by these particles depends on the velocity, in the way foreseen by the theory of relativity. In many other cases, where the dependence of the resistance upon the velocity could be detected, there was complete agreement between theory and experiment. We see once more the essential features of creative work in science: prediction of certain facts by theory and their confirmation by experiment.

This result suggests a further important generalization. A body at rest has mass but no kinetic energy, that is, energy of motion. A moving body has both mass and kinetic energy. It resists change of velocity more strongly than the resting body. It seems as though the kinetic energy of the moving body increases its resistance. If two bodies have the same rest mass, the one with the greater kinetic energy resists the action of an external force more strongly.

Imagine a box containing balls, with the box as weil

as the balls at rest in our CS. To move it, to increase its velocity, some force is required. But will the same force increase the velocity by the same amount in the same time with the balls moving about quickly and in all directions inside the box, like the molecules of a gas, with an average speed approaching that of light? A greater force will now be necessary because of the increased kinetic energy of the balls, strengthening the resistance of the box. Energy, at any rate kinetic energy, resists motion in the same way as ponderable masses. Is this also true of all kinds of energy?

The theory of relativity deduces, from its fundamental assumption, a clear and convincing answer to this question, an answer again of a quantitative character: all energy resists change of motion; all energy behaves like matter; a piece of iron weighs more when red-hot than when cool; radiation traveling through space and emitted from the sun contains energy and therefore has mass; the sun and all radiating stars lose mass by emitting radiation. This conclusion, quite general in character, is an important achievement of the theory of relativity and fits all facts upon which it has been tested.

Classical physics introduced two substances: matter and energy. The first had weight, but the second was weightless. In classical physics we had two conservation laws: one for matter, the other for energy. We have already asked whether modern physics still holds this view of two substances and the two conservation laws. The answer is: "No." According to the theory of relativity, there is no essential distinction between mass and energy. Energy has mass and mass represents energy. Instead of two conservation laws we have only one, that of mass-energy. This new view proved very successful and fruitful in the further development of physics.

How is it that this fact of energy having mass and mass representing energy remained for so long obscured? Is the weight of a piece of hot iron greater than that of a cold piece? The answer to this question is now "yes," but on p. 43 it was "no." The pages between these two answers are certainly not sufficient to cover this contradiction.

The difficulty confronting us here is of the same kind as we have met before. The variation of mass predicted by the theory of relativity is immeasurably small and cannot be detected by direct weighing on even the most sensitive scales. The proof that energy is not weightless can be gained in many very conclusive, but indirect, ways.

The reason for this lack of immediate evidence is the very small rate of exchange between matter and energy. Compared to mass, energy is like a depreciated currency compared to one of high value. An example will make this clear. The quantity of heat able to convert thirty thousand tons of water into steam would weigh about one gram! Energy was regarded as weightless for so long simply because the mass which it represents is so small.

The old energy-substance is the second victim of the theory of relativity. The first was the medium through which light waves were propagated.

The influence of the theory of relativity goes far beyond the problem from which it arose. It removes the difficulties and contradictions of the field theory;

it formulates more general mechanical laws; it replaces two conservation laws by one; it changes our classical concept of absolute time. Its validity is not restricted to one domain of physics; it forms a general framework embracing all phenomena of nature.

THE TIME-SPACE CONTINUUM

"The French revolution began in Paris on the 14th of July, 1789." In this sentence the place and time of an event are stated. Hearing this statement for the first time one who does not know what "Paris" means could be taught: it is a city on our earth situated in long. 2° East and lat. 49° North. The two numbers would then characterize the place, and "14th of July, 1789" the time, at which the event took place. In physics, much more than in history, the exact characterization of when and where an event takes place is very important, because these data form the basis for a quantitative description.

For the sake of simplicity, we considered previously only motion along a straight line. A rigid rod with an origin but no end point was our CS. Let us keep this restriction. Take different points on the rod; their positions can be characterized by one number only, by the co-ordinate of the point. To say the co-ordinate of a point is 7.586 feet means that its distance is 7.586 feet from the origin of the rod. If, on the contrary, someone gives me any number and a unit, I can always find a point on the rod corresponding to this number. We can state: a definite point on the rod corresponds to every number, and a definite number corresponds to every point. This fact is expressed by mathematicians

in the following sentence: all points on the rod form a
one-dimensional continuum. There exists a point arbitrarily near every point on the rod. We can connect
two distant points on the rod by steps as small as we
wish. Thus the arbitrary smallness of the steps connecting distant points is characteristic of the continuum.

Now another example. We have a plane, or, if you
prefer something more concrete, the surface of a rec-
tangular table. The position of a point on this table
can be characterized by two numbers and not, as before, by one. The two numbers are the distances from
two perpendicular edges of the table. Not one num-

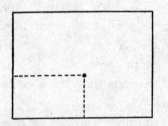

ber, but a pair of numbers corresponds to every point
on the plane; a definite point corresponds to every pair
of numbers. In other words: the plane is a *two-dimensional continuum*. There exist points arbitrarily near
every point on the plane. Two distant points can be
connected by a curve divided into steps as small as
we wish. Thus the arbitrary smallness of the steps connecting two distant points, each of which can be represented by two numbers, is again characteristic of a
two-dimensional continuum.

One more example. Imagine that you wish to regard your room as your CS. This means that you want

to describe all positions with respect to the rigid walls of the room. The position of the end point of the lamp, if the lamp is at rest, can be described by three numbers: two of them determine the distance from two perpendicular walls, and the third that from the floor or ceiling. Three definite numbers correspond to

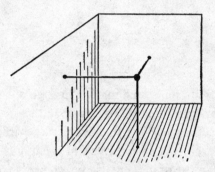

every point of the space; a definite point in space corresponds to every three numbers. This is expressed by the sentence: Our space is a *three-dimensional continuum*. There exist points very near every point of the space. Again the arbitrary smallness of the steps connecting the distant points, each of them represented by three numbers, is characteristic of a three-dimensional continuum.

But all this is scarcely physics. To return to physics, the motion of material particles must be considered. To observe and predict events in nature we must consider not only the place but also the time of physical happenings. Let us again take a very simple example.

A small stone, which can be regarded as a particle, is dropped from a tower. Imagine the tower 256 feet high. Since Galileo's time we have been able to predict the co-ordinate of the stone at any arbitrary instant after it was dropped. Here is a "timetable" describing the positions of the stone after 0, 1, 2, 3, and 4 seconds.

Time in seconds	Elevation from the ground in feet
0	256
1	240
2	192
3	112
4	0

Five events are registered in our "timetable," each represented by two numbers, the time and space co-ordinates of each event. The first event is the dropping of the stone from 256 feet above the ground at the zero second. The second event is the coincidence of the stone with our rigid rod (the tower) at 240 feet above the ground. This happens after the first second. The last event is the coincidence of the stone with the earth.

We could represent the knowledge gained from our "timetable" in a different way. We could represent the five pairs of numbers in the "timetable" as five points on a surface. Let us first establish a scale. One segment will correspond to a foot and another to a second. For example:

|← *100 Ft.* →| |← *1 Sec.* →|

We then draw two perpendicular lines, calling the horizontal one, say, the time axis and the vertical one the space axis. We see immediately that our "timetable" can be represented by five points in our time-space plane.

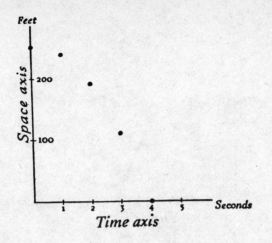

The distances of the points from the space axis represent the time co-ordinates as registered in the first column of our "timetable," and the distances from the time axis their space co-ordinates.

Exactly the same thing is expressed in two different ways: by the "timetable" and by the points on the plane. Each can be constructed from the other. The choice between these two representations is merely a matter of taste, for they are, in fact, equivalent.

Let us now go one step further. Imagine a better "timetable" giving the positions not for every second, but for, say, every hundredth or thousandth of a second. We shall then have very many points on our time-space plane. Finally, if the position is given for every instant or, as the mathematicians say, if the space co-ordinate is given as a function of time, then our set of points becomes a continuous line. Our next drawing therefore represents not just a fragment as before, but a complete knowledge of the motion.

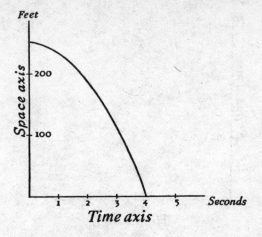

The motion along the rigid rod (the tower), the motion in a one-dimensional space, is here represented as a curve in a two-dimensional time-space continuum. To every point in our time-space continuum there corresponds a pair of numbers, one of which denotes the time, and the other the space, co-ordinate. Conversely: a definite point in our time-space plane corresponds to every pair of numbers characterizing an event. Two adjacent points represent two events, two happenings, at slightly different places and at slightly different instants.

You could argue against our representation thus: there is little sense in representing a unit of time by a segment, in combining it mechanically with the space, forming the two-dimensional continuum from the two one-dimensional continua. But you would then have to protest just as strongly against all the graphs representing, for example, the change of temperature in New York City during last summer, or against those representing the changes in the cost of living during the last few years, since the very same method is used in each of these cases. In the temperature graphs the one-dimensional temperature continuum is combined with the one-dimensional time continuum into the two-dimensional temperature-time continuum.

Let us return to the particle dropped from a 256-foot tower. Our graphic picture of motion is a useful convention since it characterizes the position of the particle at an arbitrary instant. Knowing how the particle moves, we should like to picture its motion once more. We can do this in two different ways.

We remember the picture of the particle changing its position with time in the one-dimensional space. We picture the motion as a sequence of events in the one-dimensional space continuum. We do not mix time and space, using a *dynamic* picture in which positions *change* with time.

But we can picture the same motion in a different way. We can form a *static* picture, considering the curve in the two-dimensional time-space continuum. Now the motion is represented as something which *is*, which exists in the two-dimensional time-space continuum, and not as something which changes in the one-dimensional space continuum.

Both these pictures are exactly equivalent and preferring one to the other is merely a matter of convention and taste.

Nothing that has been said here about the two pictures of the motion has anything whatever to do with the relativity theory. Both representations can be used with equal right, though classical physics favored rather the dynamic picture describing motion as happenings in space and not as existing in time-space. But the relativity theory changed this view. It was distinctly in favor of the static picture and found in this representation of motion as something existing in time-space a more convenient and more objective picture of reality. We still have to answer the question: why are these two pictures, equivalent from the point of view of classical physics, not equivalent from the point of view of the relativity theory?

The answer will be understood if two CS moving uniformly, relative to each other, are again taken into account.

According to classical physics, observers in two CS moving uniformly, relative to each other, will assign different space co-ordinates, but the same time co-ordinate, to a certain event. Thus in our example, the coincidence of the particle with the earth is characterized in our chosen CS by the time co-ordinate "4" and by the space co-ordinate "0." According to classical mechanics, the stone will still reach the earth after four seconds for an observer moving uniformly, relative to the chosen CS. But this observer will refer the distance to his CS and will, in general, connect different space co-ordinates with the event of collision, although the time co-ordinate will be the same for him and for all other observers moving uniformly, relative to each other. Classical physics knows only an "absolute" time flow for all observers. For every CS the two-dimensional continuum can be split into two one-dimensional continua: time and space. Because of the "absolute" character of time, the transition from the "static" to the "dynamic" picture of motion has an objective meaning in classical physics.

But we have already allowed ourselves to be convinced that the classical transformation must not be used in physics generally. From a practical point of view it is still good for small velocities, but not for settling fundamental physical questions.

According to the relativity theory the time of the collision of the stone with the earth will not be the same for all observers. The time co-ordinate and the space co-ordinate will be different in two CS, and the change in the time co-ordinate will be quite distinct if the relative velocity is close to that of light. The two-dimensional continuum cannot be split into two one-dimensional continua as in classical physics. We must not consider space and time separately in determining the time-space co-ordinates in another CS. The splitting of the two-dimensional continuum into two one-dimensional ones seems, from the point of view of the relativity theory, to be an arbitrary procedure without objective meaning.

It will be simple to generalize all that we have just said for the case of motion not restricted to a straight line. Indeed, not two, but four, numbers must be used to describe events in nature. Our physical space as conceived through objects and their motion has three dimensions, and positions are characterized by three numbers. The instant of an event is the fourth number. Four definite numbers correspond to every event; a definite event corresponds to any four numbers. Therefore: The world of events forms a *four-dimensional continuum*. There is nothing mysterious about this, and the last sentence is equally true for classical physics and the relativity theory. Again a difference is revealed when two CS moving relatively to each other are considered. The room is moving, and the observers inside and outside determine the time-space coordinates of the same events. Again the classical physicist splits the four-dimensional continua into the three-dimensional spaces and the one-dimensional time-continuum. The old physicist bothers only about space transformation, as time is absolute for him. He finds the splitting of the four-dimensional world-continua into space and time natural and convenient. But from the point of view of the relativity theory, time as well as space is changed by passing from one CS to another, and the Lorentz transformation considers the transformation properties of the four-dimensional time-space continuum of our four-dimensional world of events.

The world of events can be described dynamically by a picture changing in time and thrown onto the background of the three-dimensional space. But it can also be described by a static picture thrown onto the background of a four-dimensional time-space continuum. From the point of view of classical physics the two pictures, the dynamic and the static, are equivalent. But from the point of view of the relativity theory the static picture is the more convenient and the more objective.

Even in the relativity theory we can still use the dynamic picture if we prefer it. But we must remember that this division into time and space has no objective meaning since time is no longer "absolute." We shall still use the "dynamic" and not the "static" language in the following pages, bearing in mind its limitations.

GENERAL RELATIVITY

There still remains one point to be cleared up. One of the most fundamental questions has not been settled as yet: does an inertial system exist? We have learned something about the laws of nature, their invariance with respect to the Lorentz transformation, and their validity for all inertial systems moving uniformly, relative to each other. We have the laws but do not know the frame to which to refer them.

In order to be more aware of this difficulty, let us interview the classical physicist and ask him some simple questions:

"What is an inertial system?"

"It is a CS in which the laws of mechanics are valid. A body on which no external forces are acting moves uniformly in such a CS. This property thus enables us to distinguish an inertial CS from any other."

"But what does it mean to say that no forces are acting on a body?"

"It simply means that the body moves uniformly in an inertial CS."

Here we could once more put the question: "What is an inertial CS?" But since there is little hope of obtaining an answer differing from the above, let us try to gain some concrete information by changing the question:

"Is a CS rigidly connected with the earth an inertial one?"

"No, because the laws of mechanics are not rigorously valid on the earth, due to its rotation. A CS rigidly connected with the sun can be regarded for many problems as an inertial CS; but when we speak of the rotating sun, we again understand that a CS connected with it cannot be regarded as strictly inertial."

"Then what, concretely, is your inertial CS, and how is its state of motion to be chosen?"

"It is merely a useful fiction and I have no idea how to realize it. If I could only get far away from all material bodies and free myself from all external influences, my CS would then be inertial."

"But what do you mean by a CS free from all external influences?"

"I mean that the CS is inertial."

Once more we are back at our initial question!

Our interview reveals a grave difficulty in classical physics. We have laws, but do not know what frame to refer them to, and our whole physical structure seems to be built on sand.

We can approach this same difficulty from a different point of view. Try to imagine that there is only one body, forming our CS, in the entire universe. This body begins to rotate. According to classical mechanics, the physical laws for a rotating body are different from those for a non-rotating body. If the inertial principle is valid in one case it is not valid in the other. But all this sounds very suspicious. Is it permissible to consider the motion of only one body in the entire universe? By the motion of a body we always mean its change of position in relation to a second body. It is, therefore, contrary to common sense to speak about the motion of only one body. Classical mechanics and common sense disagree violently on this point. Newton's recipe is: if the inertial principle is valid, then the CS is either at rest or in uniform motion. If the inertial principle is invalid, then the body is in nonuniform motion. Thus, our verdict of motion or rest depends upon whether or not all the physical laws are applicable to a given CS.

Take two bodies, the sun and the earth, for instance. The motion we observe is again *relative*. It can be described by connecting the CS with either the earth or the sun. From this point of view, Copernicus' great achievement lies in transferring the CS from the earth to the sun. But as motion is relative and any frame of reference can be used, there seems to be no reason for favoring one CS rather than the other.

Physics again intervenes and changes our common-sense point of view. The CS connected with the sun resembles an inertial system more than that connected with the earth. The physical laws should be applied to Copernicus' CS rather than to Ptolemy's. The greatness of Copernicus' discovery can be appreciated only from the physical point of view. It illustrates the great advantage of using a CS connected rigidly with the sun for describing the motion of planets.

No absolute uniform motion exists in classical physics. If two CS are moving uniformly, relative to each other, then there is no sense in saying, "This CS is at rest and the other is moving." But if two CS are moving non-uniformly, relative to each other, then there is very good reason for saying, "This body moves and the other is at rest (or moves uniformly)." Absolute motion has here a very definite meaning. There is, at this point, a wide gulf between common sense and classical physics. The difficulties mentioned, that of an inertial system and that of absolute motion, are strictly connected with each other. Absolute motion is made possible only by the idea of an inertial system, for which the laws of nature are valid.

It may seem as though there is no way out of these difficulties, as though no physical theory can avoid them. Their root lies in the validity of the laws of nature for a special class of CS only, the inertial. The possibility of solving these difficulties depends on the answer to the following question. Can we formulate physical laws so that they are valid for all CS, not only those moving uniformly, but also those moving quite arbitrarily, relative to each other? If this can be done, our difficulties will be over. We shall then be able to apply the laws of nature to any CS. The struggle, so violent in the early days of science, between the views of Ptolemy and Copernicus would then be quite meaningless. Either CS could be used with equal justification. The two sentences, "the sun is at rest and the earth moves," or "the sun moves and the earth is at rest," would simply mean two different conventions concerning two different CS.

Could we build a real relativistic physics valid in all CS; a physics in which there would be no place for absolute, but only for relative motion? This is indeed possible!

We have at least one indication, though a very weak one, of how to build the new physics. Really relativistic physics must apply to all CS and, therefore, also to the special case of the inertial CS. We already know the laws for this inertial CS. The new general laws valid for all CS must, in the special case of the inertial system, reduce to the old, known laws.

The problem of formulating physical laws for every

CS was solved by the so-called *general relativity theory*; the previous theory, applying only to inertial systems, is called the *special relativity theory*. The two theories cannot, of course, contradict each other, since we must always include the old laws of the special relativity theory in the general laws for an inertial system. But just as the inertial CS was previously the only one for which physical laws were formulated, so now it will form the special limiting case, as all CS moving arbitrarily, relative to each other, are permissible.

This is the program for the general theory of relativity. But in sketching the way in which it was accomplished we must be even vaguer than we have been so far. New difficulties arising in the development of science force our theory to become more and more abstract. Unexpected adventures still await us. But our final aim is always a better understanding of reality. Links are added to the chain of logic connecting theory and observation. To clear the way leading from theory to experiment of unnecessary and artificial assumptions, to embrace an ever-wider region of facts, we must make the chain longer and longer. The simpler and more fundamental our assumptions become, the more intricate is our mathematical tool of reasoning; the way from theory to observation becomes longer, more subtle, and more complicated. Although it sounds paradoxical, we could say: Modern physics is simpler than the old physics and seems, therefore, more difficult and intricate. The simpler our picture of the external world and the more facts it embraces, the stronger it reflects in our minds the harmony of the universe.

Our new idea is simple: to build a physics valid for all CS. Its fulfillment brings formal complications and forces us to use mathematical tools different from those so far employed in physics. We shall show here only the connection between the fulfillment of this program and two principal problems: gravitation and geometry.

The law of inertia marks the first great advance in physics; in fact, its real beginning. It was gained by the contemplation of an idealized experiment, a body moving forever with no friction nor any other external forces acting. From this example and later from many others, we recognized the importance of the idealized experiment created by thought. Here again, idealized experiments will be discussed. Although these may sound very fantastic they will, nevertheless, help us to understand as much about relativity as is possible by our simple methods.

We had previously the idealized experiments with a uniformly moving room. Here, for a change, we shall have a falling elevator.

Imagine a great elevator at the top of a skyscraper much higher than any real one. Suddenly the cable supporting the elevator breaks, and the elevator falls freely toward the ground. Observers in the elevator are performing experiments during the fall. In describing them, we need not bother about air resistance or friction, for we may disregard their existence under our idealized conditions. One of the observers takes a handkerchief and a watch from his pocket and drops them. What happens to these two bodies? For the outside observer, who is looking through the window of the elevator, both handkerchief and watch fall toward the ground in exactly the same way, with the same acceleration. We remember that the acceleration of a falling body is quite independent of its mass and that it was this fact which revealed the equality of gravitational and inertial mass (p. 37). We also remember that the equality of the two masses, gravitational and inertial, was quite accidental from the point of view of classical mechanics and played no role in its structure. Here, however, this equality reflected in the equal acceleration of all falling bodies is essential and forms the basis of our whole argument.

Let us return to our falling handkerchief and watch; for the outside observer they are both falling with the same acceleration. But so is the elevator, with its walls, ceiling, and floor. Therefore: the distance between the two bodies and the floor will not change. For the inside observer the two bodies remain exactly where they were when he let them go. The inside observer may ignore the gravitational field, since its source lies outside his CS. He finds that no forces inside the elevator act upon the two bodies, and so they are at rest, just as if they were in an inertial CS. Strange things happen in the elevator! If the observer pushes a body in any direction, up or down for instance, it always moves uniformly so long as it does not collide with the ceiling or the floor of the elevator. Briefly speaking, the laws of classical mechanics are valid for the observer inside the elevator. All bodies behave in the way expected by the law of inertia. Our new CS rigidly connected with the freely falling elevator differs from the inertial CS in only one respect. In an inertial CS, a moving body on which no forces are acting will move uniformly forever. The inertial CS as represented in classical physics is neither limited in space nor time. The case of the observer in our elevator is, however, different. The inertial character of his CS is limited in space and time. Sooner or later the uniformly moving body will collide with the wall of the elevator, destroying the uniform motion. Sooner or later the whole elevator will collide with the earth destroying the observers and their experiments. The CS is only a "pocket edition" of a real inertial CS.

This local character of the CS is quite essential. If our imaginary elevator were to reach from the North Pole to the Equator, with the handkerchief placed over the North Pole and the watch over the Equator, then, for the outside observer, the two bodies would not have the same acceleration; they would not be at rest relative to each other. Our whole argument would fail! The dimensions of the elevator must be limited so that the equality of acceleration of all bodies relative to the outside observer may be assumed.

With this restriction, the CS takes on an inertial character for the inside observer. We can at least indicate a CS in which all the physical laws are valid, even though it is limited in time and space. If we imagine another CS, another elevator moving uniformly, relative to the one falling freely, then both these CS will be locally inertial. All laws are exactly the same in both. The transition from one to the other is given by the Lorentz transformation.

Let us see in what way both the observers, outside and inside, describe what takes place in the elevator.

The outside observer notices the motion of the elevator and of all bodies in the elevator, and finds them in agreement with Newton's gravitational law. For him, the motion is not uniform, but accelerated, because of the action of the gravitational field of the earth.

However, a generation of physicists born and brought up in the elevator would reason quite differently. They would believe themselves in possession of an inertial system and would refer all laws of nature to their elevator, stating with justification that the laws take on a specially simple form in their CS. It would be natural for them to assume their elevator at rest and their CS the inertial one.

It is impossible to settle the differences between the outside and the inside observers. Each of them could claim the right to refer all events to his CS. Both descriptions of events could be made equally consistent.

We see from this example that a consistent description of physical phenomena in two different CS is possible, even if they are not moving uniformly, relative to each other. But for such a description we must take into account gravitation, building so to speak, the "bridge" which effects a transition from one CS to the other. The gravitational field exists for the outside observer; it does not for the inside observer. Accelerated motion of the elevator in the gravitational field exists for the outside observer, rest and absence of the gravitational field for the inside observer. But the "bridge," the gravitational field, making the description in both CS possible, rests on one very important pillar: the equivalence of gravitational and inertial mass. Without this clew, unnoticed in classical mechanics, our present argument would fail completely.

Now for a somewhat different idealized experiment. There is, let us assume, an inertial CS, in which the law of inertia is valid. We have already described what happens in an elevator resting in such an inertial CS. But we now change our picture. Someone outside has fastened a rope to the elevator and is pulling, with a constant force, in the direction indicated in our drawing. It is immaterial how this is done. Since the laws of mechanics are valid in this CS, the whole elevator moves with a constant acceleration in the direction of the motion. Again we shall listen to the explanation of

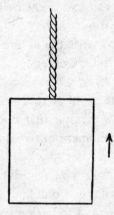

phenomena going on in the elevator and given by both the outside and inside observers.

The outside observer: My CS is an inertial one. The elevator moves with constant acceleration, because a constant force is acting. The observers inside are in absolute motion, for them the laws of mechanics are invalid. They do not find that bodies, on which no forces are acting, are at rest. If a body is left free, it soon collides with the floor of the elevator, since the floor moves upward toward the body. This happens exactly in the same way for a watch and for a handkerchief. It seems very strange to me that the observer inside the elevator must always be on the "floor" because as soon as he jumps, the floor will reach him again.

The inside observer: I do not see any reason for believing that my elevator is in absolute motion. I agree that my CS, rigidly connected with my elevator, is not really inertial, but I do not believe that it has anything to do with absolute motion. My watch, my handkerchief, and all bodies are falling because the whole elevator is in a gravitational field. I notice exactly the same kinds of motion as the man on the earth. He explains them very simply by the action of a gravitational field. The same holds good for me.

These two descriptions, one by the outside, the other by the inside, observer, are quite consistent, and there is no possibility of deciding which of them is right. We may assume either one of them for the description of phenomena in the elevator: either nonuniform motion and absence of a gravitational field with the outside observer, or rest and the presence of a gravitational field with the inside observer.

The outside observer may assume that the elevator is in "absolute" nonuniform motion. But a motion which is wiped out by the assumption of an acting gravitational field cannot be regarded as absolute motion.

There is, possibly, a way out of the ambiguity of two such different descriptions, and a decision in favor of one against the other could perhaps be made. Imagine that a light ray enters the elevator horizontally through a side window and reaches the opposite wall after a very short time. Again let us see how the path of the light would be predicted by the two observers.

The outside observer, believing in accelerated motion of the elevator, would argue: The light ray enters the window and moves horizontally, along a straight line and with a constant velocity, toward the opposite wall. But the elevator moves upward and during the time in which the light travels toward the wall, the elevator changes its position. Therefore, the ray will meet a point not exactly opposite its point of entrance, but a little below. The difference will be very slight, but it exists nevertheless, and the light ray travels, relative to the elevator, not along a straight, but along a

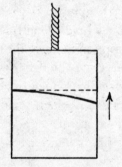

slightly curved line. The difference is due to the distance covered by the elevator during the time the ray is crossing the interior.

The inside observer, who believes in the gravitational field acting on all objects in his elevator, would say: there is no accelerated motion of the elevator, but only the action of the gravitational field. A beam of light is weightless and, therefore, will not be affected by the gravitational field. If sent in a horizontal direction, it will meet the wall at a point exactly opposite to that at which it entered.

It seems from this discussion that there is a possibility of deciding between these two opposite points of view as the phenomenon would be different for the two observers. If there is nothing illogical in either of the explanations just quoted, then our whole previous argument is destroyed, and we cannot describe all phenomena in two consistent ways, with and without a gravitational field.

But there is, fortunately, a grave fault in the reasoning of the inside observer, which saves our previous conclusion. He said: "A beam of light is weightless and, therefore, it will not be affected by the gravitational field." This cannot be right! A beam of light carries energy and energy has mass. But every inertial mass is attracted by the gravitational field as inertial and gravitational masses are equivalent. A beam of light will bend in a gravitational field exactly as a body would if thrown horizontally with a velocity equal to that of light. If the inside observer had reasoned correctly and had taken into account the bending of light rays in a gravitational field, then his results would have been exactly the same as those of an outside observer.

The gravitational field of the earth is, of course, too weak for the bending of light rays in it to be proved directly, by experiment. But the famous experiments performed during the solar eclipses show, conclusively though indirectly, the influence of a gravitational field on the path of a light ray.

It follows from these examples that there is a well-founded hope of formulating a relativistic physics. But for this we must first tackle the problem of gravitation.

We saw from the example of the elevator the consistency of the two descriptions. Nonuniform motion may, or may not, be assumed. We can eliminate "absolute" motion from our examples by a gravitational field. But then there is nothing absolute in the nonuniform motion. The gravitational field is able to wipe it out completely.

The ghosts of absolute motion and inertial CS can be expelled from physics and a new relativistic physics built. Our idealized experiments show how the problem of the general relativity theory is closely connected with that of gravitation and why the equivalence of gravitational and inertial mass is so essential for this connection. It is clear that the solution of the gravitational problem in the general theory of relativity must differ from the Newtonian one. The laws of gravitation must, just as all laws of nature, be formulated for all possible CS, whereas the laws of classical mechanics, as formulated by Newton, are valid only in inertial CS.

GEOMETRY AND EXPERIMENT

Our next example will be even more fantastic than the one with the falling elevator. We have to approach a new problem; that of a connection between the general relativity theory and geometry. Let us begin with the description of a world in which only two-dimensional and, not as in ours, three-dimensional creatures live. The movies have accustomed us to two-dimensional creatures acting on a two-dimensional screen. Now let us imagine that these shadow figures, that is, the actors on the screen, really do exist, that they have the power of thought, that they can create their own science, that for them a two-dimensional screen stands for geometrical space. These creatures are unable to imagine, in a concrete way, a three-dimensional space just as we are unable to imagine a world of four dimensions. They can deflect a straight line; they know what a circle is, but they are unable to construct a sphere, because this would mean forsaking their two-dimensional screen. We are in a similar position. We are able to deflect and curve lines and surfaces, but we can scarcely picture a deflected and curved three-dimensional space.

By living, thinking, and experimenting, our shadow figures could eventually master the knowledge of the two-dimensional Euclidean geometry. Thus, they could prove, for example, that the sum of the angles in a triangle is 180 degrees. They could construct two circles with a common center, one very small, the other large. They would find that the ratio of the circumferences of two such circles is equal to the ratio of their radii, a result again characteristic of Euclidean geometry. If the screen were infinitely great, these shadow beings would find that once having started a journey straight ahead, they would never return to their point of departure.

Let us now imagine these two-dimensional creatures living in changed conditions. Let us imagine that someone from the outside, the "third dimension," transfers them from the screen to the surface of a sphere with a very great radius. If these shadows are very small in relation to the whole surface, if they have no means of distant communication and cannot move very far, then they will not be aware of any change. The sum of angles in small triangles still amounts to 180 degrees. Two small circles with a common center still show that the ratio of their radii and circumferences are equal. A journey along a straight line never leads them back to the starting point.

But let these shadow beings, in the course of time, develop their theoretical and technical knowledge. Let them find means of communication which will enable them to cover large distances swiftly. They will then find that starting on a journey straight ahead, they ultimately return to their point of departure. "Straight ahead" means along the great circle of the sphere. They will also find that the ratio of two circles with a common center is not equal to the ratio of the radii, if one of the radii is small and the other great.

If our two-dimensional creatures are conservative, if they have learned the Euclidean geometry for genera-tions past when they could not travel far and when this geometry fitted the facts observed, they will cer-tainly make every possible effort to hold on to it, de-spite the evidence of their measurements. They could try to make physics bear the burden of these discrep-ancies. They could seek some physical reasons, say temperature differences, deforming the lines and caus-ing deviation from Euclidean geometry. But, sooner or later, they must find out that there is a much more logical and convincing way of describing these occur-rences. They will eventually understand that their world is a finite one, with different geometrical prin-ciples from those they learned. They will understand that, in spite of their inability to imagine it, their world is the two-dimensional surface of a sphere. They will soon learn new principles of geometry, which though differing from the Euclidean can, nevertheless, be for-mulated in an equally consistent and logical way for their two-dimensional world. For the new generation brought up with a knowledge of the geometry of the sphere, the old Euclidean geometry will seem more complicated and artificial since it does not fit the facts observed.

Let us return to the three-dimensional creatures of our world.

What is meant by the statement that our three-dimensional space has a Euclidean character? The meaning is that all logically proved statements of the Euclidean geometry can also be confirmed by actual experiment. We can, with the help of rigid bodies or light rays, construct objects corresponding to the idealized objects of Euclidean geometry. The edge of a ruler or a light ray corresponds to the line; the sum of the angles of a triangle built of thin rigid rods is 180 degrees; the ratio of the radii of two circles with a common center constructed from thin unbendable wire is equal to that of their circumferences. Interpreted in this way, the Euclidean geometry becomes a chapter of physics, though a very simple one.

But we can imagine that discrepancies have been discovered, for instance, that the sum of the angles of a large triangle constructed from rods, which for many reasons had to be regarded as rigid, is not 180 degrees. Since we are already used to the idea of the concrete representation of the objects of Euclidean geometry by rigid bodies, we should probably seek some physical force as the cause of such unexpected misbehavior of our rods. We should try to find the physical nature of this force and its influence on other phenomena. To save the Euclidean geometry, we should accuse the objects of not being rigid, of not exactly corresponding to those of Euclidean geometry. We should try to find a better representation of bodies behaving in the way expected by Euclidean geometry. If, however, we should not succeed in combining Euclidean geometry and physics into a simple and consistent picture, we should have to give up the idea of our space being Euclidean and seek a more convincing picture of reality under more general assumptions about the geometrical character of our space.

The necessity for this can be illustrated by an idealized experiment showing that a really relativistic physics cannot be based upon Euclidean geometry. Our argument will imply results already learned about inertial CS and the special relativity theory.

Imagine a large disk with two circles with a common center drawn on it, one very small, the other very large. The disk rotates quickly. The disk is rotating relative to an outside observer, and there is an inside observer on the disk. We further assume that the CS of the outside observer is an inertial one. The outside observer

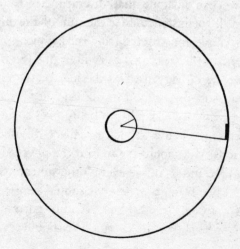

may draw, in his inertial CS, the same two circles, small and large, resting in his CS but coinciding with the circles on the rotating disk. Euclidean geometry is valid in his CS since it is inertial, so that he will find the ratio of the circumferences equal to that of the radii. But how about the observer on the disk? From the point of view of classical physics and also the special relativity theory, his CS is a forbidden one. But if we intend to find new forms for physical laws, valid in any CS, then we must treat the observer on the disk and the observer outside with equal seriousness. We, from the outside, are now watching the inside observer in his attempt to find, by measurement, the circumferences and radii on the rotating disk. He uses the same small measuring stick used by the outside observer. "The same" means either really the same, that is, handed by the outside observer to the inside, or, one of two sticks having the same length when at rest in a CS.

The inside observer on the disk begins measuring the radius and circumference of the small circle. His result must be the same as that of the outside observer. The axis on which the disk rotates passes through the center. Those parts of the disk near the center have very small velocities. If the circle is small enough we can safely apply classical mechanics and ignore the special relativity theory. This means that the stick has the same length for the outside and inside observers, and the result of these two measurements will be the same for them both. Now the observer on the disk measures the radius of the large circle. Placed on the radius, the stick moves, for the outside observer. Such a stick, however, does not contract and will have the same length for both observers since the direction of the motion is perpendicular to the stick. Thus three measurements are the same for both observers: two radii and the small circumference. But it is not so with the fourth measurement! The length of the large circumference will be different for the two observers. The stick placed on the circumference in the direction of the motion will now appear contracted to the outside observer, compared to his resting stick. The velocity is much greater than that of the inner circle, and this contraction should be taken into account. If, therefore, we apply the results of the special relativity theory, our conclusion here is: the length of the great circumference must be different if measured by the two observers. Since only one of the four lengths measured by the two observers is not the same for them both, the ratio of the two radii cannot be equal to the ratio of the two circumferences for the inside observer as it is for the outside one. This means that the observer on the disk cannot confirm the validity of Euclidean geometry in his CS.

After obtaining this result, the observer on the disk could say that he does not wish to consider CS in which Euclidean geometry is not valid. The breakdown of the Euclidean geometry is due to absolute rotation, to the fact that his CS is a bad and forbidden one. But, in arguing in this way, he rejects the principal idea of the general theory of relativity. On the other hand, if we wish to reject absolute motion and to keep up the idea of the general theory of relativity, then physics must all be built on the basis of a geometry more general than the Euclidean. There is no way of escape from this consequence if all CS are permissible.

The changes brought about by the general relativity

theory cannot be confined to space alone. In the special relativity theory we had clocks resting in every CS, having the same rhythm and synchronized, that is, showing the same time simultaneously. What happens to a clock in a noninertial CS? The idealized experiment with the disk will again be of use. The outside observer has in his inertial CS perfect clocks all having the same rhythm, all synchronized. The inside observer takes two clocks of the same kind and places one on the small inner circle and the other on the large outer circle. The clock on the inner circle has a very small velocity relative to the outside observer. We can, therefore, safely conclude that its rhythm will be the same as that of the outside clock. But the clock on the large circle has a considerable velocity, changing its rhythm compared to the clocks of the outside observer and, therefore, also compared to the clock placed on the small circle. Thus, the two rotating clocks will have different rhythms and, applying the results of the special relativity theory, we again see that in our rotating CS we can make no arrangements similar to those in an inertial CS.

To make clear what conclusions can be drawn from this and previously described idealized experiments, let us once more quote a dialogue between the old physicist O, who believes in classical physics, and the modern physicist M, who knows the general relativity theory. O is the outside observer, in the inertial CS, whereas M is on the rotating disk.

O. In your CS, Euclidean geometry is not valid. I watched your measurements and I agree that the ratio of the two circumferences is not, in your CS, equal to the ratio of the two radii. But this shows that your CS is a forbidden one. My CS, however, is of an inertial character, and I can safely apply Euclidean geometry. Your disk is in absolute motion and, from the point of view of classical physics, forms a forbidden CS, in which the laws of mechanics are not valid.

M. I do not want to hear anything about absolute motion. My CS is just as good as yours. What I noticed was your rotation relative to my disk. No one can forbid me to relate all motions to my disk.

O. But did you not feel a strange force trying to keep you away from the center of the disk? If your disk were not a rapidly rotating merry-go-round, the two things which you observed would certainly not have happened. You would not have noticed the force pushing you toward the outside nor would you have noticed that Euclidean geometry is not applicable in your CS. Are not these facts sufficient to convince you that your CS is in absolute motion?

M. Not at all! I certainly noticed the two facts you mention, but I hold a strange gravitational field acting on my disk responsible for them both. The gravitational field, being directed toward the outside of the disk, deforms my rigid rods and changes the rhythm of my clocks. The gravitational field, non-Euclidean geometry, clocks with different rhythms are, for me, all closely connected. Accepting any CS, I must at the same time assume the existence of an appropriate gravitational field with its influence upon rigid rods and clocks.

O. But are you aware of the difficulties caused by your general relativity theory? I should like to make my point clear by taking a simple nonphysical example. Imagine an idealized American town consisting of

parallel streets with parallel avenues running perpendicular to them. The distance between the streets and
also between the avenues is always the same. With
these assumptions fulfilled, the blocks are of exactly
the same size. In this way I can easily characterize the
position of any block. But such a construction would
be impossible without Euclidean geometry. Thus, for
instance, we cannot cover our whole earth with one
great ideal American town. One look at the globe will
convince you. But neither could we cover your disk
with such an "American town construction." You
claim that your rods are deformed by the gravitational
field. The fact that you could not confirm Euclid's
theorem about the equality of the ratio of radii and
circumferences shows clearly that if you carry such a
construction of streets and avenues far enough you
will, sooner or later, get into difficulties and find that it
is impossible on your disk. Your geometry on your
rotating disk resembles that on a curved surface, where,
of course, the streets-and-avenues construction is impossible on a great enough part of the surface. For a
more physical example take a plane irregularly heated
with different temperatures on different parts of the
surface. Can you, with small iron sticks expanding in
length with temperature, carry out the "parallel-per-
pendicular" construction which I have drawn below?
Of course not! Your "gravitational field" plays the
same tricks on your rods as the change of temperature
on the small iron sticks.

   M. All this does not frighten me. The street-avenue
construction is needed to determine positions of points,
with the clock to order events. The town need not be

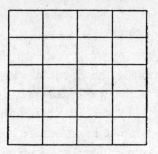

American, it could just as well be ancient European. Imagine your idealized town made of plasticine and then deformed. I can still number the blocks and recognize the streets and avenues, though these are no longer straight and equidistant. Similarly, on our earth, longitude and latitude denote the positions of points, although there is no "American town" construction.

O. But I still see a difficulty. You are forced to use your "European town structure." I agree that you can

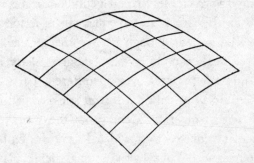

order points, or events, but this construction will muddle all measurement of distances. It will not give you the *metric properties* of space as does my construction. Take an example. I know, in my American town, that to walk ten blocks I have to cover twice the distance of five blocks. Since I know that all blocks are equal, I can immediately determine distances.

M. That is true. In my "European town" structure, I cannot measure distances immediately by the number of deformed blocks. I must know something more; I must know the geometrical properties of my surface. Just as everyone knows that from 0° to 10° longitude on the Equator is not the same distance as from 0° to 10° longitude near the North Pole. But every navigator knows how to judge the distance between two such points on our earth because he knows its geometrical properties. He can either do it by calculations based on the knowledge of spherical trigonometry, or he can do it experimentally, sailing his ship through the two distances at the same speed. In your case the whole problem is trivial, because all the streets and avenues are the same distance apart. In the case of our earth it is more complicated; the two meridians 0° and 10° meet at the earth's poles and are furthest apart on the Equator. Similarly, in my "European town structure," I must know something more than you in your "American town structure," in order to determine distances. I can gain this additional knowledge by studying the geometrical properties of my continuum in every particular case.

O. But all this only goes to show how inconvenient and complicated it is to give up the simple structure of the Euclidean geometry for the intricate scaffolding which you are bound to use. Is this really necessary?

M. I am afraid it is, if we want to apply our physics to any CS, without the mysterious inertial CS. I agree that my mathematical tool is more complicated than yours, but my physical assumptions are simpler and more natural.

The discussion has been restricted to two-dimensional continua. The point at issue in the general relativity theory is still more complicated, since it is not the two-dimensional, but the four-dimensional time-space continuum. But the ideas are the same as those sketched in the two-dimensional case. We cannot use in the general relativity theory the mechanical scaffolding of parallel, perpendicular rods and synchronized clocks, as in the special relativity theory. In an arbitrary CS we cannot determine the point and the instant at which an event happens by the use of rigid rods, rhythmical and synchronized clocks, as in the inertial CS of the special relativity theory. We can still order the events with our non-Euclidean rods and our clocks out of rhythm. But actual measurements requiring rigid rods and perfect rhythmical and synchronized clocks can be performed only in the local inertial CS. For this the whole special relativity theory is valid; but our "good" CS is only local, its inertial character being limited in space and time. Even in our arbitrary CS we can foresee the results of measurements made in the local inertial CS. But for this we must know the geometrical character of our time-space continuum.

Our idealized experiments indicate only the general character of the new relativistic physics. They show us that our fundamental problem is that of gravitation. They also show us that the general relativity theory leads to further generalization of time and space concepts.

GENERAL RELATIVITY AND ITS VERIFICATION

The general theory of relativity attempts to formulate physical laws for all CS. The fundamental problem of the theory is that of gravitation. The theory makes the first serious effort, since Newton's time, to reformulate the law of gravitation. Is this really necessary? We have already learned about the achievements of Newton's theory, about the great development of astronomy based upon his gravitational law. Newton's law still remains the basis of all astronomical calculations. But we also learned about some objections to the old theory. Newton's law is valid only in the inertial CS of classical physics, in CS defined, we remember, by the condition that the laws of mechanics must be valid in them. The force between two masses depends upon their distance from each other. The connection between force and distance is, as we know, invariant with respect to the classical transformation. But this law does not fit the frame of special relativity. The distance is not invariant with respect to the Lorentz transformation. We could try, as we did so successfully with the laws of motion, to generalize the gravitational law, to make it fit the special relativity theory, or, in other words, to formulate it so that it would be invariant with respect to the Lorentz and not to the classical transformation. But Newton's gravitational law opposed obstinately all our efforts to simplify and fit it into the scheme of the special relativity theory. Even if we succeeded in this, a further step would still be necessary: the step from the inertial CS of the special relativity theory to the arbitrary CS of the general relativity theory. On the other hand, the idealized experiments about the falling elevator show clearly that there is no chance of formulating the general relativity theory without solving the problem of gravitation. From our argument we see why the solution of the gravitational problem will differ in classical physics and general relativity.

We have tried to indicate the way leading to the general relativity theory and the reasons forcing us to change our old views once more. Without going into the formal structure of the theory, we shall characterize some features of the new gravitational theory as compared with the old. It should not be too difficult to grasp the nature of these differences in view of all that has previously been said.

1. The gravitational equations of the general relativity theory can be applied to any CS. It is merely a matter of convenience to choose any particular CS in a special case. Theoretically all CS are permissible. By ignoring the gravitation, we automatically come back to the inertial CS of the special relativity theory.

2. Newton's gravitational law connects the motion of a body here and now with the action of a body at the same time in the far distance. This is the law which formed a pattern for our whole mechanical view. But the mechanical view broke down. In Maxwell's equations we realized a new pattern for the laws of nature. Maxwell's equations are structure laws. They connect events which happen now and here with events which will happen a little later in the immediate vicinity. They are the laws describing the changes of the electromagnetic field. Our new gravitational equations are

also structure laws describing the changes of the gravitational field. Schematically speaking, we could say: the transition from Newton's gravitational law to general relativity resembles somewhat the transition from the theory of electric fluids with Coulomb's law to Maxwell's theory.

3. Our world is not Euclidean. The geometrical nature of our world is shaped by masses and their velocities. The gravitational equations of the general relativity theory try to disclose the geometrical properties of our world.

Let us suppose, for the moment, that we have succeeded in carrying out consistently the program of the general relativity theory. But are we not in danger of carrying speculation too far from reality? We know how well the old theory explains astronomical observations. Is there a possibility of constructing a bridge between the new theory and observation? Every speculation must be tested by experiment, and any results, no matter how attractive, must be rejected if they do not fit the facts. How did the new theory of gravitation stand the test of experiment? This question can be answered in one sentence: The old theory is a special limiting case of the new one. If the gravitational forces are comparatively weak, the old Newtonian law turns out to be a good approximation to the new laws of gravitation. Thus all observations which support the classical theory also support the general relativity theory. We regain the old theory from the higher level of the new one.

Even if no additional observation could be quoted in

favor of the new theory, if its explanation were only just as good as the old one, given a free choice between the two theories, we should have to decide in favor of the new one. The equations of the new theory are, from the formal point of view, more complicated, but their assumptions are, from the point of view of fundamental principles, much simpler. The two frightening ghosts, absolute time and an inertial system, have disappeared. The clew of the equivalence of gravitational and inertial mass is not overlooked. No assumption about the gravitational forces and their dependence on distance is needed. The gravitational equations have the form of structure laws, the form required of all physical laws since the great achievements of the field theory.

Some new deductions, not contained in Newton's gravitational law, can be drawn from the new gravitational laws. One, the bending of light rays in a gravitational field, has already been quoted. Two further consequences will now be mentioned.

If the old laws follow from the new one when the gravitational forces are weak, the deviations from the Newtonian law of gravitation can be expected only for comparatively strong gravitational forces. Take our solar system. The planets, our earth among them, move along elliptical paths around the sun. Mercury is the planet nearest the sun. The attraction between the sun and Mercury is stronger than that between the sun and any other planet, since the distance is smaller. If there is any hope of finding a deviation from Newton's law, the greatest chance is in the case of Mercury. It follows, from classical theory, that the path described by

Mercury is of the same kind as that of any other planet except that it is nearer the sun. According to the general relativity theory, the motion should be slightly different. Not only should Mercury travel around the sun, but the ellipse which it describes should rotate very slowly, relative to the CS connected with the sun. This rotation of the ellipse expresses the new effect of the general relativity theory. The new theory predicts the magnitude of this effect. Mercury's ellipse would perform a complete rotation in three million years! We see how small the effect is, and how hopeless it would be to seek it in the case of planets further removed from the sun.

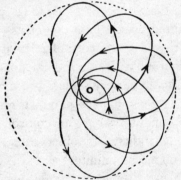

The deviation of the motion of the planet Mercury from the ellipse was known before the general relativity theory was formulated, and no explanation could be found. On the other hand, general relativity developed without any attention to this special problem. Only later was the conclusion about the rotation of the ellipse in the motion of a planet around the sun drawn from the new gravitational equations. In the case of Mercury, theory explained successfully the deviation of the motion from the Newtonian law.

But there is still another conclusion which was drawn from the general relativity theory and compared with experiment. We have already seen that a clock placed on the large circle of a rotating disk has a different rhythm from one placed on the smaller circle. Similarly, it follows from the theory of relativity that a clock placed on the sun would have a different rhythm from one placed on the earth, since the influence of the gravitational field is much stronger on the sun than on the earth.

We remarked on p. 99 that sodium, when incandescent, emits homogeneous yellow light of a definite wave-length. In this radiation the atom reveals one of its rhythms, the atom represents, so to speak, a clock and the emitted wave-length one of its rhythms. According to the general theory of relativity, the wave-length of light emitted by a sodium atom, say, placed on the sun should be very slightly greater than that of light emitted by a sodium atom on our earth.

The problem of testing the consequences of the general relativity theory by observation is an intricate one and by no means definitely settled. As we are concerned with principal ideas, we do not intend to go deeper into this matter, and only state that the verdict of experiment seems, so far, to confirm the conclusions drawn from the general relativity theory.

FIELD AND MATTER

We have seen how and why the mechanical point of view broke down. It was impossible to explain all phenomena by assuming that simple forces act between unalterable particles. Our first attempts to go beyond the mechanical view and to introduce field concepts proved most successful in the domain of electromagnetic phenomena. The structure laws for the electromagnetic field were formulated; laws connecting events very near to each other in space and time. These laws fit the frame of the special relativity theory, since they are invariant with respect to the Lorentz transformation. Later the general relativity theory formulated the gravitational laws. Again they are structure laws describing the gravitational field between material particles. It was also easy to generalize Maxwell's laws so that they could be applied to any CS, like the gravitational laws of the general relativity theory.

We have two realities: *matter and field*. There is no doubt that we cannot at present imagine the whole of physics built upon the concept of matter as the physicists of the early nineteenth century did. For the moment we accept both the concepts. Can we think of matter and field as two distinct and different realities? Given a small particle of matter we could picture in a naïve way that there is a definite surface of the particle where it ceases to exist and its gravitational field appears. In our picture, the region in which the laws of field are valid is abruptly separated from the region in which matter is present. But what are the physical criterions distinguishing matter and field? Before we learned about the relativity theory we could have tried to answer this question in the following way: matter has mass, whereas field has not. Field represents energy, matter represents mass. But we already know that such an answer is insufficient in view of the further knowledge gained. From the relativity theory we know that matter represents vast stores of energy and that energy represents matter. We cannot, in this way, distinguish qualitatively between matter and field, since the distinction between mass and energy is not a qualitative one. By far the greatest part of energy is concentrated in matter; but the field surrounding the particle also represents energy, though in an incomparably smaller quantity. We could therefore say: Matter is where the concentration of energy is great, field where the concentration of energy is small. But if this is the case, then the difference between matter and field is a quantitative rather than a qualitative one. There is no sense in regarding matter and field as two qualities quite different from each other. We cannot imagine a definite surface separating distinctly field and matter.

The same difficulty arises for the charge and its field. It seems impossible to give an obvious qualitative criterion for distinguishing between matter and field or charge and field.

Our structure laws, that is, Maxwell's laws and the gravitational laws, break down for very great concentrations of energy or, as we may say, where sources of the field, that is electric charges or matter, are present. But could we not slightly modify our equations so that they would be valid everywhere, even in regions where energy is enormously concentrated?

We cannot build physics on the basis of the matter-concept alone. But the division into matter and field is, after the recognition of the equivalence of mass and energy, something artificial and not clearly defined. Could we not reject the concept of matter and build a pure field physics? What impresses our senses as matter is really a great concentration of energy into a comparatively small space. We could regard matter as the regions in space where the field is extremely strong. In this way a new philosophical background could be created. Its final aim would be the explanation of all events in nature by structure laws valid always and everywhere. A thrown stone is, from this point of view, a changing field, where the states of greatest field intensity travel through space with the velocity of the stone. There would be no place, in our new physics, for both field and matter, field being the only reality. This new view is suggested by the great achievements of field physics, by our success in expressing the laws of electricity, magnetism, gravitation in the form of structure laws, and finally by the equivalence of mass and energy. Our ultimate problem would be to modify our field laws in such a way that they would not break down for regions in which the energy is enormously concentrated.

But we have not so far succeeded in fulfilling this program convincingly and consistently. The decision, as to whether it is possible to carry it out, belongs to the future. At present we must still assume in all our actual theoretical constructions two realities: field and matter.

Fundamental problems are still before us. We know that all matter is constructed from a few kinds of particles only. How are the various forms of matter built from these elementary particles? How do these elementary particles interact with the field? By the search for an answer to these questions new ideas have been introduced into physics, the ideas of the *quantum theory*.

WE SUMMARIZE:

A new concept appears in physics, the most important invention since Newton's time: the field. It needed great scientific imagination to realize that it is not the charges nor the particles but the field in the space between the charges and the particles which is essential for the description of physical phenomena. The field concept proves most successful and leads to the formulation of Maxwell's equations describing the structure of the electromagnetic field and governing the electric as well as the optical phenomena.

The theory of relativity arises from the field problems. The contradictions and inconsistencies of the old theories force us to ascribe new properties to the time-space continuum, to the scene of all events in our physical world.

The relativity theory develops in two steps. The first step leads to what is known as the special theory of relativity, applied only to inertial co-ordinate systems, that is, to systems in which the law of inertia, as formulated by Newton, is valid. The special theory of relativity is based on two fundamental assumptions: physical laws are the same in all co-ordinate systems moving uniformly, relative to each other; the velocity of light always has the same value. From these assumptions, fully confirmed by experiment, the properties of moving rods and clocks, their changes in length and rhythm depending on velocity, are deduced. The theory of relativity changes the laws of mechanics. The old laws are invalid if the velocity of the moving particle approaches that of light. The new laws for a moving body as reformulated by the relativity theory are splendidly confirmed by experiment. A further consequence of the (special) theory of relativity is the connection between mass and energy. Mass is energy and energy has mass. The two conservation laws of mass and energy are combined by the relativity theory into one, the conservation law of mass-energy.

The general theory of relativity gives a still deeper analysis of the time-space continuum. The validity of the theory is no longer restricted to inertial co-ordinate systems. The theory attacks the problem of gravitation and formulates new structure laws for the gravitational field. It forces us to analyze the role played by geometry in the description of the physical world. It regards the fact that gravitational and inertial mass are equal, as essential and not merely accidental, as in classical mechanics. The experimental consequences of the general relativity theory differ only slightly from those of classical mechanics. They stand the test of experiment well wherever comparison is possible. But the strength of the theory lies in its inner consistency and the simplicity of its fundamental assumptions.

The theory of relativity stresses the importance of the field concept in physics. But we have not yet succeeded in formulating a pure field physics. For the present we must still assume the existence of both: field and matter.

IV. QUANTA

Quanta

CONTINUITY—DISCONTINUITY

A MAP of New York City and the surrounding country is spread before us. We ask: which points on this map can be reached by train? After looking up these points in a railway timetable, we mark them on the map. We now change our question and ask: which points can be reached by car? If we draw lines on the map representing all the roads starting from New York, every point on these roads can, in fact, be reached by car. In both cases we have sets of points. In the first they are separated from each other and represent the different railway stations, and in the second they are the points along the lines representing the roads. Our next question is about the distance of each of these points from New York, or, to be more rigorous, from a certain spot in that city. In the first case, certain numbers correspond to the points on our map. These numbers change by irregular, but always finite, leaps and bounds. We say: the distances from New York of the places which can be reached by train change only in a *discontinuous* way. Those of the places which

249

can be reached by car, however, may change by steps as small as we wish, they can vary in a *continuous* way. The changes in distance can be made arbitrarily small in the case of a car, but not in the case of a train.

The output of a coal mine can change in a continuous way. The amount of coal produced can be decreased or increased by arbitrarily small steps. But the number of miners employed can change only discontinuously. It would be pure nonsense to say: "Since yesterday, the number of employees has increased by 3.783."

Asked about the amount of money in his pocket, a man can give a number containing only two decimals. A sum of money can change only by jumps, in a discontinuous way. In America the smallest permissible change or, as we shall call it, the "elementary quantum" for American money, is one cent. The elementary quantum for English money is one farthing, worth only half the American elementary quantum. Here we have an example of two elementary quanta whose mutual values can be compared. The ratio of their values has a definite sense since one of them is worth twice as much as the other.

We can say: some quantities can change continuously and others can change only discontinuously, by steps which cannot be further decreased. These indivisible steps are called the *elementary quanta* of the particular quantity to which they refer.

We can weigh large quantities of sand and regard its mass as continuous even though its granular structure is evident. But if the sand were to become very precious and the scales used very sensitive, we should have to consider the fact that the mass always changes by a multiple number of one grain. The mass of this one grain would be our elementary quantum. From this example we see how the discontinuous character of a quantity, so far regarded as continuous, can be detected by increasing the precision of our measurements.

If we had to characterize the principal idea of the quantum theory in one sentence, we could say: *it must be assumed that some physical quantities so far regarded as continuous are composed of elementary quanta.*

The region of facts covered by the quantum theory is tremendously great. These facts have been disclosed by the highly developed technique of modern experiment. As we can neither show nor describe even the basic experiments, we shall frequently have to quote their results dogmatically. Our aim is to explain the principal underlying ideas only.

ELEMENTARY QUANTA OF MATTER AND ELECTRICITY

In the picture of matter drawn by the kinetic theory, all elements are built of molecules. Take the simplest case of the lightest element, that is hydrogen. On p. 62 we saw how the study of Brownian motions led to the determination of the mass of one hydrogen molecule. Its value is:

0.000 000 000 000 000 000 000 0033 grams.

This means that mass is discontinuous. The mass of a portion of hydrogen can change only by a whole number of small steps each corresponding to the mass of one hydrogen molecule. But chemical processes show that the hydrogen molecule can be broken up into two parts, or, in other words, that the hydrogen molecule is composed of two atoms. In chemical processes it is the atom and not the molecule which plays the role of an elementary quantum. Dividing the above number by two, we find the mass of a hydrogen atom. This is about

0.000 000 000 000 000 000 000 0017 grams.

Mass is a discontinuous quantity. But, of course, we need not bother about this when determining weight. Even the most sensitive scales are far from attaining the degree of precision by which the discontinuity in mass variation could be detected.

Let us return to a well-known fact. A wire is connected with the source of a current. The current is flowing through the wire from higher to lower potential. We remember that many experimental facts were explained by the simple theory of electric fluids flowing through the wire. We also remember (p. 79) that the decision as to whether the positive fluid flows from higher to lower potential, or the negative fluid flows from lower to higher potential, was merely a matter of convention. For the moment we disregard all the further progress resulting from the field concepts. Even when thinking in the simple terms of electric fluids, there still remain some questions to be settled. As the name "fluid" suggests, electricity was regarded, in the early days, as a continuous quantity. The amount of charge could be changed, according to these old views, by arbitrarily small steps. There was no need to assume elementary electric quanta. The achievements of the kinetic theory of matter prepared us for a new question: do elementary quanta of electric fluids exist? The other question to be settled is: does the current consist of a flow of positive, negative or perhaps of both fluids?

The idea of all the experiments answering these questions is to tear the electric fluid from the wire, to let it travel through empty space, to deprive it of any association with matter and then to investigate its properties, which must appear most clearly under these conditions. Many experiments of this kind were performed in the late nineteenth century. Before explaining the idea of these experimental arrangements, at least in one case, we shall quote the results. The electric fluid flowing through the wire is a negative one, directed, therefore, from lower to higher potential. Had we known this from the start, when the theory of electric fluids was first formed, we should certainly have interchanged the words, and called the electricity of the rubber rod positive, that of the glass rod negative. It would then have been more convenient to regard the flowing fluid as the positive one. Since our first guess was wrong we now have to put up with the inconvenience. The next important question is whether the structure of this negative fluid is "granular," whether or not it is composed of electric quanta. Again a number of independent experiments show that there is no doubt as to the existence of an elementary quantum of this negative electricity. The negative electric fluid is constructed of grains, just as the beach is composed of grains of sand, or a house built of bricks. This result was formulated most clearly by J. J. Thomson, about forty years ago. The elementary quanta of negative electricity are called *electrons*. Thus every negative electric charge is composed of a multitude of elementary charges represented by electrons. The negative charge can, like mass, vary only discontinuously. The elementary electric charge is, however, so small that in many investigations it is equally possible and sometimes even more convenient to regard it as a continuous quantity. Thus the atomic and electron theories introduce into science discontinuous physical quantities which can vary only by jumps.

Imagine two parallel metal plates in some place from which all air has been extracted. One of the plates has a positive, the other a negative charge. A positive test charge brought between the two plates will be repelled by the positively charged and attracted by the negatively charged plate. Thus the lines of force of the electric field will be directed from the positively to the negatively charged plate. A force acting on a negatively charged test body would have the opposite direction. If the plates are sufficiently large, the lines of force between them will be equally dense everywhere; it is immaterial where the test body is placed, the force and, therefore, the density of the lines of force will be

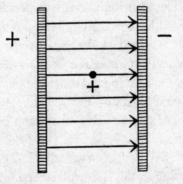

the same. Electrons brought somewhere between the plates would behave like raindrops in the gravitational field of the earth, moving parallel to each other from the negatively to the positively charged plate. There are many known experimental arrangements for bringing a shower of electrons into such a field which directs them all in the same way. One of the simplest is to bring a heated wire between the charged plates. Such a heated wire emits electrons which are afterwards directed by the lines of force of the external field. For instance, radio tubes, familiar to everyone, are based on this principle.

Many very ingenious experiments have been performed on a beam of electrons. The changes of their path in different electric and magnetic external fields have been investigated. It has even been possible to isolate a single electron and to determine its elementary charge and its mass, that is, its inertial resistance to the action of an external force. Here we shall only quote the value of the mass of an electron. It turned out to be about *two thousand times smaller* than the mass of a hydrogen atom. Thus the mass of a hydrogen atom, small as it is, appears great in comparison with the mass of an electron. From the point of view of a consistent field theory, the whole mass, that is, the whole energy, of an electron is the energy of its field; the bulk of its strength is within a very small sphere, and away from the "center" of the electron it is weak.

We said before that the atom of any element is its smallest elementary quantum. This statement was believed for a very long time. Now, however, it is no longer believed! Science has formed a new view showing the limitations of the old one. There is scarcely any statement in physics more firmly founded on facts than the one about the complex structure of the atom.

First came the realization that the electron, the elementary quantum of the negative electric fluid, is also one of the components of the atom, one of the elementary bricks from which all matter is built. The previously quoted example of a heated wire emitting electrons is only one of the numerous instances of the extraction of these particles from matter. This result closely connecting the problem of the structure of matter with that of electricity follows, beyond any doubt, from very many independent experimental facts.

It is comparatively easy to extract from an atom some of the electrons from which it is composed. This can be done by heat, as in our example of a heated wire, or in a different way, such as by bombarding atoms with other electrons.

Suppose a thin, red-hot, metal wire is inserted into rarefied hydrogen. The wire will emit electrons in all directions. Under the action of a foreign electric field a given velocity will be imparted to them. An electron increases its velocity just as a stone falling in the gravitational field. By this method we can obtain a beam of electrons rushing along with a definite speed in a definite direction. Nowadays, we can reach velocities comparable to that of light by submitting electrons to the action of very strong fields. What happens, then, when a beam of electrons of a definite velocity impinges on the molecules of rarefied hydrogen? The impact of a sufficiently speedy electron will not only disrupt the hydrogen molecule into its two atoms but will also extract an electron from one of the atoms.

Let us accept the fact that electrons are constituents of matter. Then, an atom from which an electron has been torn out cannot be electrically neutral. If it was previously neutral, then it cannot be so now, since it is poorer by one elementary charge. That which remains must have a positive charge. Furthermore, since the mass of an electron is so much smaller than that of the lightest atom, we can safely conclude that by far the greater part of the mass of the atom is not represented by electrons but by the remainder of the elementary particles which are much heavier than the electrons. We call this heavy part of the atom its *nucleus*.

Modern experimental physics has developed methods of breaking up the nucleus of the atom, of changing atoms of one element into those of another, and of extracting from the nucleus the heavy elementary particles of which it is built. This chapter of physics, known as "nuclear physics," to which Rutherford contributed so much, is, from the experimental point of view, the most interesting. But a theory, simple in its fundamental ideas and connecting the rich variety of facts in the domain of nuclear physics, is still lacking. Since, in these pages, we are interested only in general physical ideas, we shall omit this chapter in spite of its great importance in modern physics.

THE QUANTA OF LIGHT

Let us consider a wall built along the seashore. The waves from the sea continually impinge on the wall, wash away some of its surface, and retreat, leaving the way clear for the incoming waves. The mass of the wall decreases and we can ask how much is washed away in, say, one year. But now let us picture a different process. We want to diminish the mass of the wall by the same amount as previously but in a different way. We shoot at the wall and split it at the places where the bullets hit. The mass of the wall will be decreased and we can well imagine that the same reduction in mass is achieved in both cases. But from the appearance of the wall we could easily detect whether the continuous sea wave or the discontinuous shower of bullets has been acting. It will be helpful in understanding the phenomena which we are about to describe, to bear in mind the difference between sea waves and a shower of bullets.

We said, previously, that a heated wire emits electrons. Here we shall introduce another way of extracting electrons from metal. Homogeneous light, such as violet light, which is, as we know, light of a definite wave-length, is impinging on a metal surface. The light extracts electrons from the metal. The electrons are torn from the metal and a shower of them speeds along with a certain velocity. From the point of view of the energy principle we can say: the energy of light is partially transformed into the kinetic energy of expelled electrons. Modern experimental technique enables us to register these electron-bullets, to determine their velocity and thus their energy. This extraction of electrons by light falling upon metal is called the *photoelectric effect*.

Our starting point was the action of a homogeneous light wave, with some definite intensity. As in every experiment, we must now change our arrangements to see whether this will have any influence on the observed effect.

Let us begin by changing the intensity of the homogeneous violet light falling on the metal plate and note to what extent the energy of the emitted electrons depends upon the intensity of the light. Let us try to find the answer by reasoning instead of by experiment. We could argue: in the photoelectric effect a certain definite portion of the energy of radiation is transformed into energy of motion of the electrons. If we again illuminate the metal with light of the same wave-length but from a more powerful source, then the energy of the emitted electrons should be greater, since the radiation is richer in energy. We should, therefore, expect the velocity of the emitted electrons to increase if the intensity of the light increases. But experiment again contradicts our prediction. Once more we see that the laws of nature are not as we should like them to be. We have come upon one of the experiments which, contradicting our predictions, breaks the theory on which they were based. The actual experimental result is, from the point of view of the wave theory, astonishing. The observed electrons all have the same speed, the same energy, which does not change when the intensity of the light is increased.

This experimental result could not be predicted by the wave theory. Here again a new theory arises from the conflict between the old theory and experiment.

Let us be deliberately unjust to the wave theory of light, forgetting its great achievements, its splendid explanation of the bending of light around very small obstacles. With our attention focused on the photoelectric effect, let us demand from the theory an adequate explanation of this effect. Obviously, we cannot deduce from the wave theory the independence of the energy of electrons from the intensity of light by which they have been extracted from the metal plate. We shall, therefore, try another theory. We remember that Newton's corpuscular theory, explaining many of the observed phenomena of light, failed to account for the bending of light, which we are now deliberately disregarding. In Newton's time the concept of energy did not exist. Light corpuscles were, according to him, weightless; each color preserved its own substance character. Later, when the concept of energy was created and it was recognized that light carries energy, no one thought of applying these concepts to the corpuscular theory of light. Newton's theory was dead and, until our own century, its revival was not taken seriously.

To keep the principal idea of Newton's theory, we must assume that homogeneous light is composed of energy-grains and replace the old light corpuscles by light quanta, which we shall call *photons*, small portions of energy, traveling through empty space with the velocity of light. The revival of Newton's theory in this new form leads to the *quantum theory of light*. Not only matter and electric charge, but also energy of radiation has a granular structure, i.e., is built up of light quanta. In addition to quanta of matter and quanta of electricity there are also quanta of energy.

The idea of energy quanta was first introduced by Planck at the beginning of this century in order to explain some effects much more complicated than the photoelectric effect. But the photo-effect shows most clearly and simply the necessity for changing our old concepts.

It is at once evident that this quantum theory of light explains the photoelectric effect. A shower of photons is falling on a metal plate. The action between radiation and matter consists here of very many single processes in which a photon impinges on the atom and tears out an electron. These single processes are all alike and the extracted electron will have the same energy in every case. We also understand that increasing the intensity of the light means, in our new language, increasing the number of falling photons. In this case, a different number of electrons would be thrown out of the metal plate, but the energy of any single one would not change. Thus we see that this theory is in perfect agreement with observation.

What will happen if a beam of homogeneous light of a different color, say, red instead of violet, falls on the metal surface? Let us leave experiment to answer this question. The energy of the extracted electrons must be measured and compared with the energy of electrons thrown out by violet light. The energy of the electron extracted by red light turns out to be smaller than the energy of the electron extracted by violet light. This means that the energy of the light quanta is different for different colors. The photons belonging to the color red have half the energy of those belonging to the color violet. Or, more rigorously: the energy of a light quantum belonging to a homogeneous color decreases proportionally as the wave-length increases. There is an essential difference between quanta of energy and quanta of electricity. Light quanta differ for every wave-length, whereas quanta of electricity are always the same. If we were to use one of our previous analogies, we should compare light quanta to the smallest monetary quanta, differing in each country.

Let us continue to discard the wave theory of light and assume that the structure of light is granular and is formed by light quanta, that is, photons speeding through space with the velocity of light. Thus, in our new picture, light is a shower of photons, and the photon is the elementary quantum of light energy. If, however, the wave theory is discarded, the concept of a wave-length disappears. What new concept takes its place? The energy of the light quanta! Statements expressed in the terminology of the wave theory can be translated into statements of the quantum theory of radiation. For example:

TERMINOLOGY OF THE WAVE THEORY	TERMINOLOGY OF THE QUANTUM THEORY
Homogeneous light has a definite wave-length. The wave-length of the red end of the spectrum is twice that of the violet end.	Homogeneous light contains photons of a definite energy. The energy of the photon for the red end of the spectrum is half that of the violet end.

The state of affairs can be summarized in the following way: there are phenomena which can be explained by the quantum theory but not by the wave theory. Photo-effect furnishes an example, though other phenomena of this kind are known. There are phenomena which can be explained by the wave theory but not by the quantum theory. The bending of light around obstacles is a typical example. Finally, there are phenomena, such as the rectilinear propagation of light, which can be equally well explained by the quantum and the wave theory of light.

But what is light really? Is it a wave or a shower of photons? Once before we put a similar question when we asked: is light a wave or a shower of light corpuscles? At that time there was every reason for discarding the corpuscular theory of light and accepting the wave theory, which covered all phenomena. Now, however, the problem is much more complicated. There seems no likelihood of forming a consistent description of the phenomena of light by a choice of only one of the two possible languages. It seems as though we must use sometimes the one theory and sometimes the other, while at times we may use either. We are faced with a new kind of difficulty. We have two contradictory pictures of reality; separately neither of them fully explains the phenomena of light, but together they do!

How is it possible to combine these two pictures? How can we understand these two utterly different aspects of light? It is not easy to account for this new difficulty. Again we are faced with a fundamental problem.

For the moment let us accept the photon theory of light and try, by its help, to understand the facts so far explained by the wave theory. In this way we shall stress the difficulties which make the two theories appear, at first sight, irreconcilable.

We remember: a beam of homogeneous light passing through a pinhole gives light and dark rings (p. 112 ff) How is it possible to understand this phenomena by the help of the quantum theory of light, disregarding the wave theory? A photon passes through the hole. We could expect the screen to appear light if the photon passes through and dark if it does not. Instead, we find light and dark rings. We could try to account for it as follows: perhaps there is some interaction between the rim of the hole and the photon which is responsible for the appearance of the diffraction rings. This sentence can, of course, hardly be regarded as an explanation. At best, it outlines a program for an explanation holding out at least some hope of a future understanding of diffraction by interaction between matter and photons.

But even this feeble hope is dashed by our previous discussion of another experimental arrangement. Let us take two pinholes. Homogeneous light passing through the two holes gives light and dark stripes on the screen. How is this effect to be understood from the point of view of the quantum theory of light? We could argue: a photon passes through either one of the two pinholes. If a photon of homogeneous light represents an elementary light particle, we can hardly imagine its division and its passage through the two holes. But then the effect should be exactly as in the first case, light and dark rings and not light and dark stripes. How is it possible then that the presence of another pinhole completely changes the effect? Apparently the hole through which the photon does not pass, even though it may be at a fair distance, changes the rings into stripes! If the photon behaves like a corpuscle in classical physics it must pass through one of the two holes. But in this case, the phenomena of diffraction seem quite incomprehensible.

Science forces us to create new ideas, new theories. Their aim is to break down the wall of contradictions which frequently blocks the way of scientific progress. All the essential ideas in science were born in a dramatic conflict between reality and our attempts at understanding. Here again is a problem for the solution of which new principles are needed. Before we try to account for the attempts of modern physics to explain the contradiction between the quantum and the wave aspects of light, we shall show that exactly the same difficulty appears when dealing with quanta of matter instead of quanta of light.

We already know that all matter is built of only a few kinds of particles. Electrons were the first elementary particles of matter to be discovered. But electrons are also the elementary quanta of negative electricity. We learned furthermore that some phenomena force us to assume that light is composed of elementary light quanta, differing for different wave-lengths. Before proceeding we must discuss some physical phenomena in which matter as well as radiation plays an essential role.

The sun emits radiation which can be split into its components by a prism. The continuous spectrum of the sun can thus be obtained. Every wave-length between the two ends of the visible spectrum is represented. Let us take another example. It was previously mentioned that sodium when incandescent emits homogeneous light, light of one color or one wave-length. If incandescent sodium is placed before the prism we see only one yellow line. In general, if a radiating body is placed before the prism, then the light it emits is split up into its components, revealing the spectrum characteristic of the emitting body.

The discharge of electricity in a tube containing gas produces a source of light such as seen in the neon tubes used for luminous advertisements. Suppose such a tube is placed before a spectroscope. The spectroscope is an instrument which acts like a prism, but with much greater accuracy and sensitiveness; it splits light into its components, that is, it analyzes it. Light from the sun, seen through a spectroscope, gives a continuous spectrum; all wave-lengths are represented in it. If, however, the source of light is a gas through which a current of electricity passes, the spectrum is of a different character. Instead of the continuous, multi-colored design of the sun's spectrum, bright, separated stripes appear on a continuous dark background. Every stripe, if it is very narrow, corresponds to a definite color or, in the language of the wave theory, to a definite wavelength. For example, if twenty lines are visible in the spectrum, each of them will be designated by one of twenty numbers expressing the corresponding wavelength. The vapors of the various elements possess different systems of lines, and thus different combinations of numbers designating the wave-lengths composing the emitted light spectrum. No two elements have identical systems of stripes in their characteristic spectra, just as no two persons have exactly identical fingerprints. As a catalogue of these lines was worked out by physicists, the existence of laws gradually became evident, and it was possible to replace some of the columns of seemingly disconnected numbers expressing the length of the various waves by one simple mathematical formula.

All that has just been said can now be translated into the photon language. The stripes correspond to certain definite wave-lengths or, in other words, to photons with a definite energy. Luminous gases do not, therefore, emit photons with all possible energies, but only those characteristic of the substance. Reality again limits the wealth of possibilities.

Atoms of a particular element, say, hydrogen, can emit only photons with definite energies. Only the emission of definite energy quanta is permissible, all others being prohibited. Imagine, for the sake of simplicity, that some element emits only one line, that is, photons of a quite definite energy. The atom is richer in energy before the emission and poorer afterwards. From the energy principle it must follow that the *energy level* of an atom is higher before emission and lower afterwards, and that the difference between the two levels must be equal to the energy of the emitted photon. Thus the fact that an atom of a certain element emits radiation of one wave-length only, that is photons of a definite energy only, could be expressed differently: only two energy levels are permissible in an atom of this element and the emission of a photon corresponds to the transition of the atom from the higher to the lower energy level.

But more than one line appears in the spectra of the elements, as a rule. The photons emitted correspond to many energies and not to one only. Or, in other words, we must assume that many energy levels are allowed in an atom and that the emission of a photon corresponds to the transition of the atom from a higher energy level to a lower one. But it is essential that not every energy level should be permitted, since not every wave-length, not every photon-energy, appears in the spectra of an element. Instead of saying that some definite lines, some definite wave-lengths, belong to the spectrum of every atom, we can say that every atom has some definite energy levels, and that the emission of light quanta is associated with the transition of the atom from one energy level to another. The energy levels are, as a rule, not continuous but discontinuous. Again we see that the possibilities are restricted by reality.

It was Bohr who showed for the first time why just these and no other lines appear in the spectra. His theory, formulated twenty-five years ago, draws a picture of the atom from which, at any rate in simple cases, the spectra of the elements can be calculated and the apparently dull and unrelated numbers are suddenly made coherent in the light of the theory.

Bohr's theory forms an intermediate step toward a deeper and more general theory, called the wave or quantum mechanics. It is our aim in these last pages to characterize the principal ideas of this theory. Before doing so, we must mention one more theoretical and experimental result of a more special character.

Our visible spectrum begins with a certain wavelength for the violet color and ends with a certain wave-length for the red color. Or, in other words, the energies of the photons in the visible spectrum are always enclosed within the limits formed by the photon energies of the violet and red lights. This limitation is, of course, only a property of the human eye. If the difference in energy of some of the energy levels is sufficiently great, then an *ultraviolet* photon will be sent out, giving a line beyond the visible spectrum. Its presence cannot be detected by the naked eye; a photographic plate must be used.

X rays are also composed of photons of a much greater energy than those of visible light, or in other words, their wave-lengths are much smaller, thousands of times smaller in fact, than those of visible light.

But is it possible to determine such small wave-lengths experimentally? It was difficult enough to do so for ordinary light. We had to have small obstacles or small apertures. Two pinholes very near to each other, showing diffraction for ordinary light, would have to be many thousands of times smaller and closer together to show diffraction for X rays.

How then can we measure the wave-lengths of these rays? Nature herself comes to our aid.

A crystal is a conglomeration of atoms arranged at very short distances from each other on a perfectly regular plan. Our drawing shows a simple model of the

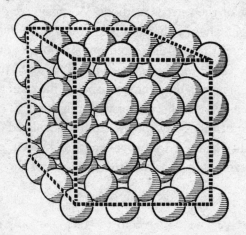

structure of a crystal. Instead of minute apertures, there are extremely small obstacles formed by the atoms of the element, arranged very close to each other in absolutely regular order. The distances between the atoms, as found from the theory of the crystal structure, are so small that they might be expected to show the effect of diffraction for X rays. Experiment proved that it is, in fact, possible to diffract the X-ray wave by means of these closely packed obstacles disposed in the regular three-dimensional arrangement occurring in a crystal.

Suppose that a beam of X rays falls upon a crystal and, after passing through it, is recorded on a photographic plate. The plate then shows the diffraction pattern. Various methods have been used to study the X-ray spectra, to deduce data concerning the wavelength from the diffraction pattern. What has been said here in a few words would fill volumes if all theoretical and experimental details were set forth. In Plate III we give only one diffraction pattern obtained by one of the various methods. We again see the dark and light rings so characteristic of the wave theory. In the center the non-diffracted ray is visible. If the crystal were not brought between the X rays and the photographic plate, only the light spot in the center would be seen. From photographs of this kind the wavelengths of the X-ray spectra can be calculated and, on the other hand, if the wave-length is known, conclusions can be drawn about the structure of the crystal.

THE WAVES OF MATTER

How can we understand the fact that only certain characteristic wave-lengths appear in the spectra of the elements?

It has often happened in physics that an essential advance was achieved by carrying out a consistent analogy between apparently unrelated phenomena. In these pages we have often seen how ideas created and developed in one branch of science were afterwards successfully applied to another. The development of

PLATE III

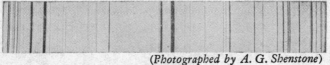

(Photographed by A. G. Shenstone)

Spectral lines.

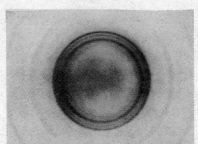

(Photographed by Lastowiecki and Gregor)

Diffraction of X rays.

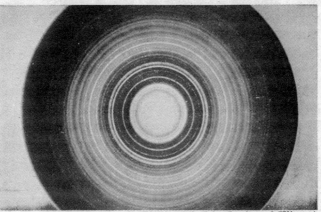

(Photographed by Loria and Klinger)

Diffraction of electronic waves.

the mechanical and field views gives many examples of this kind. The association of solved problems with those unsolved may throw new light on our difficulties by suggesting new ideas. It is easy to find a superficial analogy which really expresses nothing. But to discover some essential common features, hidden beneath a surface of external differences, to form, on this basis, a new successful theory, is important creative work. The development of the so-called wave mechanics, begun by de Broglie and Schrödinger, less than fifteen years ago, is a typical example of the achievement of a successful theory by means of a deep and fortunate analogy.

Our starting point is a classical example having nothing to do with modern physics. We take in our hand the end of a very long flexible rubber tube, or a very long spring, and try to move it rhythmically up and down, so that the end oscillates. Then, as we have seen in many other examples, a wave is created by the oscillation which spreads through the tube with a cer-

tain velocity. If we imagine an infinitely long tube, then the portions of waves, once started, will pursue their endless journey without interference.

Now another case. The two ends of the same tube are fastened. If preferred, a violin string may be used. What happens now if a wave is created at one end of the rubber tube or cord? The wave begins its journey as in the previous example, but it is soon reflected by the other end of the tube. We now have two waves:

one created by oscillation, the other by reflection; they travel in opposite directions and interfere with each other. It would not be difficult to trace the interference of the two waves and discover the one wave resulting from their superposition; it is called the *standing wave*. The two words "standing" and "wave" seem to contradict each other; their combination is, nevertheless, justified by the result of the superposition of the two waves.

The simplest example of a standing wave is the motion of a cord with the two ends fixed, an up-and-down motion, as shown in our drawing. This motion is the

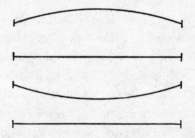

result of one wave lying on the other when the two are traveling in opposite directions. The characteristic feature of this motion is: only the two end points are at rest. They are called *nodes*. The wave stands, so to speak, between the two nodes, all points of the cord reaching simultaneously the maxima and minima of their deviation.

But this is only the simplest kind of a standing wave. There are others. For example, a standing wave can have three nodes, one at each end and one in the center. In this case three points are always at rest. A glance at the drawings shows that here the wave-length is half as great as the one with two nodes. Similarly, standing

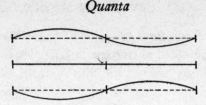

waves can have four, five, and more nodes. The wave-length in each case will depend on the number of

nodes. This number can only be an integer and can change only by jumps. The sentence, "the number of nodes in a standing wave is 3.576," is pure nonsense. Thus the wave-length can only change discontinuously. Here, in this most classical problem, we recognize the familiar features of the quantum theory. The standing wave produced by a violin player is, in fact, still more complicated, being a mixture of very many waves with two, three, four, five, and more nodes and, therefore, a mixture of several wave-lengths. Physics can analyze such a mixture into the simple standing waves from which it is composed. Or, using our previous terminology, we could say that the oscillating string has its spectrum, just as an element emitting radiation. And, in the same way as for the spectrum of an element, only certain wave-lengths are allowed, all others being prohibited.

We have thus discovered some similarity between the oscillating cord and the atom emitting radiation. Strange as this analogy may seem, let us draw further conclusions from it and try to proceed with the comparison, once having chosen it. The atoms of every element are composed of elementary particles, the heavier constituting the nucleus, and the lighter the electrons. Such a system of particles behaves like a small acoustical instrument in which standing waves are produced.

Yet the standing wave is the result of interference between two or, generally, even more moving waves. If there is some truth in our analogy, a still simpler arrangement than that of the atom should correspond to a spreading wave. What is the simplest arrangement? In our material world, nothing can be simpler than an electron, an elementary particle, on which no forces are acting, that is, an electron at rest or in uniform motion. We could guess a further link in the chain of our analogy: electron moving uniformly → waves of a definite length. This was de Broglie's new and courageous idea.

It was previously shown that there are phenomena in which light reveals its wave-like character and others in which light reveals its corpuscular character. After becoming used to the idea that light is a wave, we found, to our astonishment, that in some cases, for instance in the photoelectric effect, it behaves like a shower of photons. Now we have just the opposite state of affairs for electrons. We accustomed ourselves to the idea that electrons are particles, elementary quanta of electricity and matter. Their charge and mass were investigated. If there is any truth in de Broglie's idea, then there must be some phenomena in which matter reveals its wave-like character. At first, this conclusion, reached by following the acoustical analogy, seems strange and incomprehensible. How can a moving corpuscle have anything to do with a wave? But this is not the first time we have faced a difficulty of this kind in physics. We met the same problem in the domain of light phenomena.

Fundamental ideas play the most essential role in forming a physical theory. Books on physics are full of complicated mathematical formulae. But thought and ideas, not formulae, are the beginning of every physical theory. The ideas must later take the mathematical form of a quantitative theory, to make possible the comparison with experiment. This can be explained by the example of the problem with which we are now dealing. The principal guess is that the uniformly moving electron will behave, in some phenomena, like a wave. Assume that an electron or a shower of electrons, provided they all have the same velocity, is moving uniformly. The mass, charge, and velocity of each individual electron is known. If we wish to associate in some way a wave concept with a uniformly moving electron or electrons, our next question must be: what is the wave-length? This is a quantitative question and a more or less quantitative theory must be built up to answer it. This is indeed a simple matter. The mathematical simplicity of de Broglie's work, providing an answer to this question, is most astonishing. At the time his work was done, the mathematical technique of other physical theories was very subtle and complicated, comparatively speaking. The mathematics dealing with the problem of waves of matter is extremely simple and elementary but the fundamental ideas are deep and far-reaching.

Previously, in the case of light waves and photons, it was shown that every statement formulated in the wave language can be translated into the language of photons or light corpuscles. The same is true for electronic waves. For uniformly moving electrons, the corpuscular language is already known. But every statement expressed in the corpuscular language can be translated into the wave language, just as in the case of photons. Two clews laid down the rules of translation. The analogy between light waves and electronic waves or photons and electrons is one clew. We try to use the same method of translation for matter as for light. The special relativity theory furnished the other clew. The laws of nature must be invariant with respect to the Lorentz and not to the classical transformation. These two clews together determine the wave-length corresponding to a moving electron. It follows from the theory that an electron moving with a velocity of, say, 10,000 miles per second, has a wave-length which can be easily calculated, and which turns out to lie in the same region as the X-ray wave-lengths. Thus we conclude further that if the wave character of matter can be detected, it should be done experimentally in an analogous way to that of X rays.

Imagine an electron beam moving uniformly with a given velocity, or, to use the wave terminology, a homogeneous electronic wave, and assume that it falls on a very thin crystal, playing the part of a diffraction grating. The distances between the diffracting obstacles in the crystal are so small that diffraction for X rays can be produced. One might expect a similar effect for electronic waves with the same order of wave-length. A photographic plate would register this diffraction of electronic waves passing through the thin layer of crystal. Indeed, the experiment produces what is undoubtedly one of the great achievements of the theory: the phenomenon of diffraction for electronic waves. The similarity between the diffraction of an electronic wave and that of an X ray is particularly marked as seen from a comparison of the patterns in Plate III. We know that such pictures enable us to determine the wave-lengths of X rays. The same holds good for electronic waves. The diffraction pattern gives the length of a wave of matter and the perfect quantitative agreement between theory and experiment confirms the chain of our argument splendidly.

Our previous difficulties are broadened and deepened by this result. This can be made clear by an example similar to the one given for a light wave. An electron shot at a very small hole will bend like a light wave. Light and dark rings appear on the photographic plate. There may be some hope of explaining this phenomenon by the interaction between the electron and the rim, though such an explanation does not seem to be very promising. But what about the two pinholes? Stripes appear instead of rings. How is it possible that the presence of the other hole completely changes the effect? The electron is indivisible and can, it would seem, pass through only one of the two holes. How could an electron passing through a hole possibly know that another hole has been made some distance away?

We asked before: what is light? Is it a shower of corpuscles or a wave? We now ask: what is matter, what is an electron? Is it a particle or a wave? The electron behaves like a particle when moving in an external electric or magnetic field. It behaves like a wave when diffracted by a crystal. With the elementary quanta of matter we came across the same difficulty that we met with in the light quanta. One of the most fundamental questions raised by recent advance in science is how to reconcile the two contradictory views of matter and wave. It is one of those fundamental difficulties which, once formulated, must lead, in the long run, to scientific progress. Physics has tried to solve this problem. The future must decide whether the solution suggested by modern physics is enduring or temporary.

PROBABILITY WAVES

If, according to classical mechanics, we know the position and velocity of a given material point and also what external forces are acting, we can predict, from the mechanical laws, the whole of its future path. The sentence: "The material point has such-and-such position and velocity at such-and-such an instant," has a definite meaning in classical mechanics. If this statement were to lose its sense, our argument (p. 30) about foretelling the future path would fail.

In the early nineteenth century, scientists wanted to reduce all physics to simple forces acting on material particles that have definite positions and velocities at any instant. Let us recall how we described motion when discussing mechanics at the beginning of our journey through the realm of physical problems. We drew points along a definite path showing the exact positions of the body at certain instants and then tangent vectors showing the direction and magnitude of the velocities. This was both simple and convincing. But it cannot be repeated for our elementary quanta of matter, that is electrons, or for quanta of energy, that is photons. We cannot picture the journey of a photon or electron in the way we imagined motion in classical mechanics. The example of the two pinholes shows this clearly. Electron and photon seem to pass through the two holes. It is thus impossible to explain the effect by picturing the path of an electron or a photon in the old classical way.

We must, of course, assume the presence of elementary actions, such as the passing of electrons or photons through the holes. The existence of elementary quanta of matter and energy cannot be doubted. But the elementary laws certainly cannot be formulated by specifying positions and velocities at any instant in the simple manner of classical mechanics.

Let us, therefore, try something different. Let us continually repeat the same elementary processes. One after the other, the electrons are sent in the direction of the pinholes. The word "electron" is used here for the sake of definiteness; our argument is also valid for photons.

The same experiment is repeated over and over again in exactly the same way; the electrons all have the same velocity and move in the direction of the two pinholes. It need hardly be mentioned that this is an idealized experiment which cannot be carried out in reality but may well be imagined. We cannot shoot out single photons or electrons at given instants, like bullets from a gun.

The outcome of repeated experiments must again be dark and light rings for one hole and dark and light stripes for two. But there is one essential difference. In the case of one individual electron, the experimental result was incomprehensible. It is more easily understood when the experiment is repeated many times. We can now say: light stripes appear where many electrons fall. The stripes become darker at the place where fewer electrons are falling. A completely dark spot means that there are no electrons. We are not, of course, allowed to assume that all the electrons pass through one of the holes. If this were so it could not make the slightest difference whether or not the other is covered. But we already know that covering the second hole does make a difference. Since one particle is indivisible we cannot imagine that it passes through both the holes. The fact that the experiment was repeated many times points to another way out. Some of the electrons may pass through the first hole and others through the second. We do not know why individual electrons choose particular holes, but the net result of repeated experiments must be that both pinholes participate in transmitting the electrons from the source to the screen. If we state only what happens to the crowd of elecrons when the experiment is repeated, not bothering about the behavior of individual particles, the difference between the ringed and the striped pictures becomes comprehensible. By the discussion of a sequence of experiments a new idea was born, that of a crowd with the individuals behaving in an unpredictable way. We cannot foretell the course of one single electron, but we can predict that, in the net result, the light and dark stripes will appear on the screen.

Let us leave quantum physics for the moment.

We have seen in classical physics that if we know the position and velocity of a material point at a certain instant and the forces acting upon it, we can predict its future path. We also saw how the mechanical point of view was applied to the kinetic theory of matter. But in this theory a new idea arose from our reasoning. It will be helpful in understanding later arguments to grasp this idea thoroughly.

There is a vessel containing gas. In attempting to trace the motion of every particle one would have to commence by finding the initial states, that is, the initial positions and velocities of all the particles. Even if this were possible, it would take more than a human lifetime to set down the result on paper, owing to the enormous number of particles which would have to be considered. If one then tried to employ the known methods of classical mechanics for calculating the final positions of the particles, the difficulties would be insurmountable. In principle, it is possible to use the method applied for the motion of planets, but in practice this is useless and must give way to the *method of statistics*. This method dispenses with any exact knowledge of initial states. We know less about the system at any given moment and are thus less able to say anything about its past or future. We become indifferent to the fate of the individual gas particles. Our problem is of a different nature. For example: we do not ask, "What is the speed of every particle at this moment?" But we may ask: "How many particles have a speed between 1000 and 1100 feet per second?" We care nothing for individuals. What we seek to determine are average values typifying the whole aggregation. It is clear that there can be some point in a statistical method of reasoning only when the system consists of a large number of individuals.

By applying the statistical method we cannot foretell the behavior of an individual in a crowd. We can only foretell the chance, *the probability*, that it will behave in some particular manner. If our statistical laws tell us that one-third of the particles have a speed between 1000 and 1100 feet per second, it means that by repeating our observations for many particles, we shall really obtain this average, or in other words, that the probability of finding a particle within this limit is equal to one-third.

Similarly, to know the birth rate of a great community does not mean knowing whether any particular family is blessed with a child. It means a knowledge of statistical results in which the contributing personalities play no role.

By observing the registration plates of a great many cars we can soon discover that one-third of their numbers are divisible by three. But we cannot foretell whether the car which will pass in the next moment will have this property. Statistical laws can be applied only to big aggregations, but not to their individual members.

We can now return to our quantum problem.

The laws of quantum physics are of a statistical character. This means: they concern not one single system but an aggregation of identical systems; they cannot be verified by measurement of one individual, but only by a series of repeated measurements.

Radioactive disintegration is one of the many events for which quantum physics tries to formulate laws governing the spontaneous transmutation from one element to another. We know, for example, that in 1600 years half of one gram of radium will disintegrate, and half will remain. We can foretell approximately how many atoms will disintegrate during the next half-hour, but we cannot say, even in our theoretical descriptions, why just these particular atoms are doomed. According to our present knowledge, we have no power to designate the individual atoms condemned to disintegration. The fate of an atom does not depend on its age. There is not the slightest trace of a law governing their individual behavior. Only statistical laws can be formulated, laws governing large aggregations of atoms.

Take another example. The luminous gas of some element placed before a spectroscope shows lines of definite wave-length. The appearance of a discontinuous set of definite wave-lengths is characteristic of the atomic phenomena in which the existence of elementary quanta is revealed. But there is still another aspect of this problem. Some of the spectrum lines are very distinct, others are fainter. A distinct line means that a comparatively large number of photons belonging to this particular wave-length are emitted; a faint line means that a comparatively small number of photons belonging to this wave-length are emitted. Theory again gives us statements of a statistical nature only. Every line corresponds to a transition from higher to lower energy level. Theory tells us only about the probability of each of these possible transitions, but nothing about the actual transition of an individual atom. The theory works splendidly because all these phenomena involve large aggregations and not single individuals.

It seems that the new quantum physics resembles somewhat the kinetic theory of matter, since both are of a statistical nature and both refer to great aggregations. But this is not so! In this analogy an understanding not only of the similarities but also of the differences is most important. The similarity between the kinetic theory of matter and quantum physics lies chiefly in their statistical character. But what are the differences?

If we wish to know how many men and women over the age of twenty live in a city, we must get every citizen to fill out a form under the headings: "male," "female," and "age." Provided every answer is correct, we can obtain, by counting and segregating them, a result of a statistical nature. The individual names and addresses on the forms are of no account. Our statistical view is gained by the knowledge of individual cases. Similarly, in the kinetic theory of matter, we have statistical laws governing the aggregation, gained on the basis of individual laws.

But in quantum physics the state of affairs is entirely different. Here the statistical laws are given immediately. The individual laws are discarded. In the example of a photon or an electron and two pinholes we have seen that we cannot describe the possible motion of elementary particles in space and time as we did in classical physics. Quantum physics abandons individual laws of elementary particles and states *directly* the statistical laws governing aggregations. It is impossible, on the basis of quantum physics, to describe positions and velocities of an elementary particle or to predict its future path as in classical physics. Quantum physics deals only with aggregations, and its laws are for crowds and not for individuals.

It is hard necessity and not speculation or a desire for novelty which forces us to change the old classical view. The difficulties of applying the old view have been outlined for one instance only, that of diffraction phenomena. But many others, equally convincing, could be quoted. Changes of view are continually forced upon us by our attempts to understand reality. But it always remains for the future to decide whether we chose the only possible way out and whether or not a better solution of our difficulties could have been found.

We have had to forsake the description of individual cases as objective happenings in space and time; we have had to introduce laws of a statistical nature. These are the chief characteristics of modern quantum physics.

Previously, when introducing new physical realities, such as the electromagnetic and gravitational field, we tried to indicate in general terms the characteristic features of the equations through which the ideas have been mathematically formulated. We shall now do the same with quantum physics, referring only very briefly to the work of Bohr, De Broglie, Schrödinger, Heisenberg, Dirac and Born.

Let us consider the case of one electron. The electron may be under the influence of an arbitrary foreign electromagnetic field, or free from all external influences. It may move, for instance, in the field of an atomic nucleus or it may diffract on a crystal. Quantum physics teaches us how to formulate the mathematical equations for any of these problems.

We have already recognized the similarity between an oscillating cord, the membrane of a drum, a wind instrument, or any other acoustical instrument on the one hand, and a radiating atom on the other. There is also some similarity between the mathematical equations governing the acoustical problem and those governing the problem of quantum physics. But again the physical interpretation of the quantities determined in these two cases is quite different. The physical quantities describing the oscillating cord and the radiating atom have quite a different meaning, despite some formal likeness in the equations. In the case of the cord, we ask about the deviation of an arbitrary point from its normal position at an arbitrary moment. Knowing the form of the oscillating cord at a given instant, we know everything we wish. The deviation from the normal can thus be calculated for any other moment from the mathematical equations for the oscillating cord. The fact that some definite deviation from the normal position corresponds to every point of the cord is expressed more rigorously as follows: for any instant, the deviation from the normal value is a *function* of the co-ordinates of the cord. All points of the cord form a one-dimensional continuum, and the deviation is a function defined in this one-dimensional continuum, to be calculated from the equations of the oscillating cord.

Analogously, in the case of an electron a certain function is determined for any point in space and for any moment. We shall call this function the *probability wave*. In our analogy the probability wave corresponds to the deviation from the normal position in the acoustical problem. The probability wave is, at a given instant, a function of a three-dimensional continuum, whereas, in the case of the cord the deviation was, at a given moment, a function of the one-dimensional continuum. The probability wave forms the catalogue of our knowledge of the quantum system under consideration and will enable us to answer all sensible statistical questions concerning this system. It does not tell us the position and velocity of the electron at any moment because such a question has no sense in quantum physics. But it will tell us the probability of meeting the electron on a particular spot, or where we have the greatest chance of meeting an electron. The result does not refer to one, but to many repeated measurements. Thus the equations of quantum physics determine the probability wave just as Maxwell's equations determine the electromagnetic field and the gravitational equations determine the gravitational field. The laws of quantum physics are again structure laws. But the meaning of physical concepts determined by these equations of quantum physics is much more abstract than in the case of electromagnetic and gravitational fields; they provide only the mathematical means of answering questions of a statistical nature.

So far we have considered the electron in some external field. If it were not the electron, the smallest possible charge, but some respectable charge containing billions of electrons, we could disregard the whole quantum theory and treat the problem according to our old pre-quantum physics. Speaking of currents in a wire, of charged conductors, of electromagnetic waves, we can apply our old simple physics contained in Maxwell's equations. But we cannot do this when speaking of the photoelectric effect, intensity of spectral lines, radioactivity, diffraction of electronic waves and many other phenomena in which the quantum character of matter and energy is revealed. We must then, so to speak, go one floor higher. Whereas in classical physics we spoke of positions and velocities of one particle, we must now consider probability waves, in a three-dimensional continuum corresponding to this one-particle problem.

Quantum physics gives its own prescription for treating a problem if we have previously been taught how to treat an analogous problem from the point of view of classical physics.

For one elementary particle, electron or photon, we have probability waves in a three-dimensional continuum, characterizing the statistical behavior of the system if the experiments are often repeated. But what about the case of not one but two interacting particles, for instance, two electrons, electron and photon, or electron and nucleus? We cannot treat them separately and describe each of them through a probability wave in three dimensions, just because of their mutual interaction. Indeed, it is not very difficult to guess how to describe in quantum physics a system composed of two interacting particles. We have to descend one floor, to return for a moment to classical physics. The position of two material points in space, at any moment, is characterized by six numbers, three for each of the points. All possible positions of the two material points form a six-dimensional continuum and not a three-dimensional one as in the case of one point. If we now again ascend one floor, to quantum physics, we shall have probability waves in a six-dimensional continuum and not in a three-dimensional continuum as in the case of one particle. Similarly, for three, four, and more particles the probability waves will be functions in a continuum of nine, twelve, and more dimensions.

This shows clearly that the probability waves are

more abstract than the electromagnetic and gravitational field existing and spreading in our three-dimensional space. The continuum of many dimensions forms the background for the probability waves, and only for one particle does the number of dimensions equal that of physical space. The only physical significance of the probability wave is that it enables us to answer sensible statistical questions in the case of many particles as well as of one. Thus, for instance, for one electron we could ask about the probability of meeting an electron in some particular spot. For two particles our question could be: what is the probability of meeting the two particles at two definite spots at a given instant?

Our first step away from classical physics was abandoning the description of individual cases as objective events in space and time. We were forced to apply the statistical method provided by the probability waves. Once having chosen this way, we are obliged to go further toward abstraction. Probability waves in many dimensions corresponding to the many-particle problem must be introduced.

Let us, for the sake of briefness, call everything except quantum physics, classical physics. Classical and quantum physics differ radically. Classical physics aims at a description of objects existing in space, and the formulation of laws governing their changes in time. But the phenomena revealing the particle and wave nature of matter and radiation, the apparently statistical character of elementary events such as radioactive disintegration, diffraction, emission of spectral lines, and many others, forced us to give up this view. Quantum physics does not aim at the description of individual objects in space and their changes in time. There is no place in quantum physics for statements such as: "This object is so-and-so, has this-and-this property." Instead we have statements of this kind: "There is such-and-such a probability that the individual object is so-and-so and has this-and-this property." There is no place in quantum physics for laws governing the changes in time of the individual object. Instead, we have laws governing the changes in time of the probability. Only this fundamental change, brought into physics by the quantum theory, made possible an adequate explanation of the apparently discontinuous and statistical character of events in the realm of phenomena in which the elementary quanta of matter and radiation reveal their existence.

Yet new, still more difficult problems arise which have not been definitely settled as yet. We shall mention only some of these unsolved problems. Science is not and will never be a closed book. Every important advance brings new questions. Every development reveals, in the long run, new and deeper difficulties.

We already know that in the simple case of one or many particles we can rise from the classical to the quantum description, from the objective description of events in space and time to probability waves. But we remember the all-important field concept in classical physics. How can we describe interaction between elementary quanta of matter and field? If a probability wave in thirty dimensions is needed for the quantum description of ten particles, then a probability wave with an infinite number of dimensions would be needed for the quantum description of a field. The transition from the classical field concept to the corresponding problem of probability waves in quantum physics is a very difficult step. Ascending one floor is here no easy task and all attempts so far made to solve the problem must be regarded as unsatisfactory. There is also one other fundamental problem. In all our arguments about the transition from classical physics to quantum physics we used the old prerelativistic description in which space and time are treated differently. If, however, we try to begin from the classical description as proposed by the relativity theory, then our ascent to the quantum problem seems much more complicated. This is another problem tackled by modern physics, but still far from a complete and satisfactory solution. There is still a further difficulty in forming a consistent physics for heavy particles, constituting the nuclei. In spite of the many experimental data and the many attempts to throw light on the nuclear problem, we are still in the dark about some of the most fundamental questions in this domain.

There is no doubt that quantum physics explained a very rich variety of facts, achieving, for the most part, splendid agreement between theory and observation. The new quantum physics removes us still further from the old mechanical view, and a retreat to the former position seems, more than ever, unlikely. But there is also no doubt that quantum physics must still be based on the two concepts: matter and field. It is, in this sense, a dualistic theory and does not bring our old problem of reducing everything to the field concept even one step nearer realization.

Will the further development be along the line chosen in quantum physics, or is it more likely that new revolutionary ideas will be introduced into physics? Will the road of advance again make a sharp turn, as it has so often done in the past?

During the last few years all the difficulties of quantum physics have been concentrated around a few principal points. Physics awaits their solution impatiently. But there is no way of foreseeing when and where the clarification of these difficulties will be brought about.

PHYSICS AND REALITY

What are the general conclusions which can be drawn from the development of physics indicated here in a broad outline representing only the most fundamental ideas?

Science is not just a collection of laws, a catalogue of unrelated facts. It is a creation of the human mind, with its freely invented ideas and concepts. Physical theories try to form a picture of reality and to establish its connection with the wide world of sense impressions. Thus the only justification for our mental structures is whether and in what way our theories form such a link.

We have seen new realities created by the advance of physics. But this chain of creation can be traced back far beyond the starting point of physics. One of the most primitive concepts is that of an object. The concepts of a tree, a horse, any material body, are creations gained on the basis of experience, though the impressions from which they arise are primitive in comparison with the world of physical phenomena. A cat teasing a mouse also creates, by thought, its own primitive reality. The fact that the cat reacts in a similar way toward any mouse it meets shows that it forms concepts and theories which are its guide through its own world of sense impressions.

"Three trees" is something different from "two trees." Again "two trees" is different from "two stones." The concepts of the pure numbers 2, 3, 4 ..., freed from the objects from which they arose, are creations of the thinking mind which describe the reality of our world.

The psychological subjective feeling of time enables us to order our impressions, to state that one event precedes another. But to connect every instant of time with a number, by the use of a clock, to regard time as a one-dimensional continuum, is already an invention. So also are the concepts of Euclidean and non-Euclidean geometry, and our space understood as a three-dimensional continuum.

Physics really began with the invention of mass, force, and an inertial system. These concepts are all free inventions. They led to the formulation of the mechanical point of view. For the physicist of the early nineteenth century, the reality of our outer world consisted of particles with simple forces acting between them and depending only on the distance. He tried to retain as long as possible his belief that he would succeed in explaining all events in nature by these fundamental concepts of reality. The difficulties connected with the deflection of the magnetic needle, the difficulties connected with the structure of the ether, induced us to create a more subtle reality. The important invention of the electromagnetic field appears. A courageous scientific imagination was needed to realize fully that not the behavior of bodies, but the behavior of something between them. that is, the field, may be essential for ordering and understanding events.

Later developments both destroyed old concepts and created new ones. Absolute time and the inertial co-ordinate system were abandoned by the relativity theory. The background for all events was no longer the one-dimensional time and the three-dimensional space continuum, but the four-dimensional time-space continuum, another free invention, with new transformation properties. The inertial co-ordinate system was no longer needed. Every co-ordinate system is equally suited for the description of events in nature.

The quantum theory again created new and essential features of our reality. Discontinuity replaced continuity. Instead of laws governing individuals, probability laws appeared.

The reality created by modern physics is, indeed, far removed from the reality of the early days. But the aim of every physical theory still remains the same.

With the help of physical theories we try to find our way through the maze of observed facts, to order and understand the world of our sense impressions. We want the observed facts to follow logically from our concept of reality. Without the belief that it is possible to grasp the reality with our theoretical constructions, without the belief in the inner harmony of our world, there could be no science. This belief is and always will remain the fundamental motive for all scientific creation. Throughout all our efforts, in every dramatic struggle between old and new views, we recognize the eternal longing for understanding, the ever-firm belief in the harmony of our world, continually strengthened by the increasing obstacles to comprehension.

WE SUMMARIZE:

Again the rich variety of facts in the realm of atomic phenomena forces us to invent new physical concepts. Matter has a granular structure; it is composed of elementary particles, the elementary quanta of matter. Thus, the electric charge has a granular structure and—most important from the point of view of the quantum theory—so has energy. Photons are the energy quanta of which light is composed.

Is light a wave or a shower of photons? Is a beam of electrons a shower of elementary particles or a wave? These fundamental questions are forced upon physics by experiment. In seeking to answer them we have to abandon the description of atomic events as happenings in space and time, we have to retreat still further from the old mechanical view. Quantum physics formulates laws governing crowds and not individuals. Not properties but probabilities are described, not laws disclosing the future of systems are formulated, but laws governing the changes in time of the probabilities and relating to great congregations of individuals.

Index